AF880014

About IFPRI

The International Food Policy Research Institute (IFPRI), a CGIAR research center established in 1975, provides research-based policy solutions to sustainably reduce poverty and end hunger and malnutrition. IFPRI's strategic research aims to foster a climate-resilient and sustainable food supply; promote healthy diets and nutrition for all; build inclusive and efficient markets, trade systems, and food industries; transform agricultural and rural economies; and strengthen institutions and governance. Gender is integrated in all the Institute's work. Partnerships, communications, capacity strengthening, and data and knowledge management are essential components to translate IFPRI's research from action to impact. The Institute's regional and country programs play a critical role in responding to demand for food policy research and in delivering holistic support for country-led development. IFPRI collaborates with partners around the world.

www.ifpri.org

About CGIAR

CGIAR is a global research partnership for a food-secure future. CGIAR science is dedicated to reducing poverty, enhancing food and nutrition security, and improving natural resources and ecosystem services. Its research is carried out by 15 CGIAR Centers in close collaboration with hundreds of partners, including national and regional research institutes, civil society organizations, academia, development organizations, and the private sector.

www.cgiar.org

IFPRI

GLOBAL FOOD POLICY REPORT

2021

TRANSFORMING FOOD SYSTEMS AFTER COVID-19

A Peer-Reviewed Publication

RECOMMENDED CITATION: International Food Policy Research Institute. 2021. *2021 Global Food Policy Report: Transforming Food Systems after COVID-19.* Washington, DC: International Food Policy Research Institute. https://doi.org/10.2499/9780896293991

International Food Policy Research Institute
1201 Eye Street, NW
Washington, DC 20005-3915 USA
www.ifpri.org

ISBN: 978-0-89629-399-1
ISSN: 2329-2873
DOI: https://doi.org/10.2499/9780896293991
Cataloging-in-Publication Data is available from the Library of Congress.

Photo credits
COVER: Reuters/Alamy Stock Photo.
CHAPTER IMAGES: p. 6 Ugaraj Yathavam/Shutterstock.com; p. 18 chinahbzyg/Shutterstock.com, AS photostudio/Shutterstock.com, Atul Loke/Panos Pictures, Tommy Trenchard/Panos Pictures, p. 19 Tommy Trenchard/Panos Pictures; p. 24 Adam Dean/Panos Pictures; p. 36 Agarianna76/Shutterstock.com; p. 44 jeep2499/Shutterstock.com; p. 54 Tom Pilston/Panos Pictures; p. 64 Majority World CIC/Alamy Stock Photo; p. 74 Lucas Ledwaba/iStockphoto.

BOOK LAYOUT: Jason Chow
EDITORIAL MANAGER: Pamela Stedman-Edwards

Contents

Foreword

This is a crucial moment for the world's food systems. Hunger has been on the rise since 2015, and 3 billion people cannot afford healthy diets. At the same time, food systems are placing unsustainable demands on the world's water and energy resources and contributing a hefty share of greenhouse gas emissions. All these trends were well underway long before COVID-19.

2020, which brought the pandemic, was a year of crisis. And the terrible loss and disruption experienced worldwide will continue in many places through this year and even beyond. Increased poverty, food insecurity, malnutrition, and unemployment have pushed the Sustainable Development Goals further out of reach for many countries, and shone a harsh light on the disparities in our food systems. From these crises, however, have emerged many lessons. Foremost among them is that transforming our food systems is a matter of utmost urgency.

2021 is a year of urgency but also of hope. Vaccines are being distributed, and the health and economic shocks of the pandemic have stimulated creativity and reforms in the private and public sectors. The experience has sparked a willingness to think beyond traditional perspectives – economic, technological, and political. 2021 is also the year of global summits on food systems, climate, and nutrition. Together, this creates an unusual opportunity for the world to choose radical change.

IFPRI is contributing evidence-based inputs for these critical global policy discussions and decisions, drawn from its large set of analytical tools, data, and regional coverage. This year's *Global Food Policy Report* examines what we have learned about the deficiencies in current food systems, the changes that are needed for system transformation, and what COVID-19 has taught us. It offers lessons that can help put the world on the path to food system transformation for greater resilience, inclusion, efficiency, sustainability, and nutrition. IFPRI's analytical work during the pandemic – conducted through both new initiatives and reconfigured ongoing research – illuminates the impacts of the crisis in numerous countries and considers how policy can best address such shocks while also helping to transform food systems for the future.

We hope that this report will support transformation by contributing to the discussions at upcoming global events, and to the many national and local policy discussions and reforms that will be essential to purposeful transformation. We look forward to working together to address the fundamental changes needed for a better future.

JOHAN SWINNEN
Director General

Acknowledgments

The *2021 Global Food Policy Report* was prepared under the overall leadership of Johan Swinnen and John McDermott and a core team comprising Pamela Stedman-Edwards, Sivan Yosef, and Jamed Falik.

Text and data contributions were made by Kibrom Abay, Akhter Ahmed, Kamiljon Akramov, Kwaw Andam, Channing Arndt, Clemens Breisinger, Inge Brouwer, Samuel Benin, Emily Cho, Kevin Chen, Eugenio Díaz-Bonilla, Dalia Elsabbagh, Aulo Gelli, Ugo Gentilini, Hosam Ibrahim, Jarilkasin Ilyasov, Ahmed Kamaly, Neha Kumar, John McDermott, Ruth Meinzen-Dick, Nichola Naylor, Valeria Piñeiro, Agnes Quisumbing, Mariam Raouf, Thomas Reardon, Shahidur Rashid, Danielle Resnick, Claudia Ringler, Roman Romashkin, Christopher Rue, Marie Ruel, Sara Shapleigh, Johan Swinnen, John Ulimwengu, Rob Vos, Abdul Wajid Rana, Sivan Yosef, Yue Zhan, and Wei Zhang.

Production of the report was led by Pamela Stedman-Edwards. Jason Chow was responsible for design and layout. Editorial assistance was provided by Tracy Brown and John Whitehead.

We would like to thank IFPRI's donors, including all those who supported IFPRI's research through their contributions for the CGIAR Fund: https://cgiar.org.funders/.

CHAPTER 1

Beyond the Pandemic
Transforming Food Systems after COVID-19

JOHAN SWINNEN, JOHN McDERMOTT, AND SIVAN YOSEF

Johan Swinnen is director general of the International Food Policy Research Institute (IFPRI), Washington, DC. **John McDermott** is director of the CGIAR Research Program on Agriculture for Nutrition and Health, IFPRI, Washington, DC. **Sivan Yosef** is a senior program manager in the Director General's Office, IFPRI, Washington, DC.

KEY MESSAGES

- Before the onset of the coronavirus pandemic, our food systems already faced serious challenges in achieving equitable access to healthy, nutritious food for all; environmental sustainability; and resilience to shocks. COVID-19 has put the world further behind in reaching the UN Sustainable Development Goals (SDGs).
- COVID-19 caused widespread loss of livelihoods and incomes, threatening the food security, health, and nutrition of poor and marginalized people around the world. Countries implemented a variety of measures to mitigate these impacts, including expanded social protection; but some impacts will be long-lasting.
- Food system transformation must be pursued to regain this lost ground and achieve the SDGs by 2030.
- Yet the pandemic and associated policy responses exposed weaknesses and inequalities within food systems, including among different world regions, rural and urban communities, rich and poor populations, and disadvantaged groups such as women.
- Some food systems and sectors were more resilient than others, depending on their structure, governance, and roles of the public and private sector.
- 2020 offered lessons, innovations, and opportunities that can help make food systems more resilient to future shocks and more inclusive, efficient, sustainable, and healthy.

RECOMMENDATIONS

- Seize the opportunities opened by the pandemic – including growing momentum and lessons learned – to transform food systems to be resilient, healthy, efficient, sustainable, and inclusive.
- Use global events planned for 2021 – including UNFSS, COP26, and the Nutrition for Growth Summit – to put food system transformation prominently on the development agenda.
- Increase resilience for all food system actors through actions that limit the frequency and severity of shocks,

improve communities' ability to anticipate shocks, and build capacity to absorb shocks. This will require better access to finance; flexible social safety nets; competitive markets and trade channels; and investment in rural services, infrastructure, and R&D for improving food production systems.

- Promote the expansion and flexibility of social protection policies to protect vulnerable populations in times of economic, health, or environmental crises.
- Improve access to infrastructure and markets, especially through provision of digital services for market and farming information, education, government interactions, financial transactions, and logistics to reduce inequality and facilitate resilience.
- Seek innovative means of financing food system transformation, including through policies influencing consumer spending and private sector expenditures and profits, support for impact investment, and repurposing of public funding.

The year 2020 was unprecedented in many ways. For rich and poor countries alike, the coronavirus pandemic and the associated policy responses brought a widespread health calamity, economic hardship, severe disruptions to services, and previously unimaginable restrictions on movement. Many poor and vulnerable people have faced serious threats to their immediate food security, health, and nutrition. For the many countries that cannot access or administer vaccines quickly, the pandemic will be prolonged, with worrisome consequences for people's long-term prospects as a result of lost livelihoods, malnutrition, missed education, and depleted assets. The crisis has also highlighted and often accentuated the weaknesses and inequalities already present in our food systems, health systems, and economic systems that leave the poor and vulnerable at risk. A year into the pandemic, loss of incomes, increased food insecurity, and reduced access to healthy diets make it clear that food systems must play a central role in putting us on track to achieve the Sustainable Development Goals (SDGs) by 2030. Food systems need to be transformed to meet those goals, to better prepare us for the next shock, and to benefit the world's poor and vulnerable people and the planet.

Paradoxically, by upending our world, 2020 also offered a wide array of lessons, innovations, and opportunities that can transform our food systems not just to make them more resilient but also to make them more inclusive, efficient, sustainable, and healthy. Although income losses caused dramatic declines in food security and nutrition and increases in poverty, food supply systems proved surprisingly resilient – albeit with large differences across food commodities and regions. Many countries rapidly introduced measures to secure the flow of food products, and governments expanded social safety net programs in new ways to ensure food security. Private sector innovations introduced along food supply chains by both large companies and small and medium enterprises (SMEs) helped to overcome constraints such as lockdowns; they also stimulated investments in technologies and partnerships to keep food supply chains moving. Also importantly, in the face of the pandemic, food systems often proved able to respond rapidly and flexibly. As a result, the willingness and momentum needed to change them for the better notably increased in 2020.

The post-pandemic world thus affords us a unique opportunity to fundamentally transform food systems. The fate of billions of people, many of whom have been pushed back into poverty, food insecurity, and malnutrition, depends on quick and bold action.

A WORLD OFF TRACK

Prior to the pandemic, our food systems already faced serious challenges in achieving equitable access to healthy, nutritious food for all, environmental sustainability, and resilience to shocks. After decades of improvement, the number of hungry people in the world had been rising again for several years, largely as a result of economic slowdowns, conflicts, and extreme weather events that contributed to food crises in many low- and lower-middle-income countries.[1]

Much of the world is not on track to achieve international targets set for the next decade, including the World Health Assembly targets for 2025 and many of the SGDs. Many countries were already off track for SDG 2: Zero Hunger by 2030 before the pandemic worsened the situation.[2] Our food systems have also failed to make sufficient progress against malnutrition.[3] Many poor countries now face the triple burden of malnutrition, that is, the coexistence of undernutrition, micronutrient deficiencies, and overweight and obesity, and more than 3 billion people worldwide cannot afford a healthy diet (Chapter 3).[4] In addition, agricultural production and other activities along the food value chain are stressing our finite natural resources, biodiversity, and the environment (Chapter 4). Globally, agrifood systems consume more than 30 percent of energy and produce more than 20 percent of greenhouse gases (GhGs).[5] Climate change constitutes one of the greatest threats to our and future generations.

Food system transformation is the clearest path to overcoming the massive challenges ahead. What do we need from our food systems? Ideal food systems have five critical attributes (Figure 1).[6] They are **efficient**, providing incentives and removing hurdles for the private sector – from large businesses to smallholder farmers – to deliver efficiencies all along the food supply chain, including in crop production, infrastructure, food storage and transportation, and food consumption. They contribute to global **health**, producing affordable, nutritious foods, boosting demand for them among consumers, and guarding food safety. They are **inclusive** of smallholder farmers and marginalized groups such as women, youth, the landless, refugees, and displaced people, helping them to build decent livelihoods and to benefit as consumers and participants in decision-making. They are environmentally **sustainable**, using technological innovations, regulations, and local collective governance approaches to conserve and protect natural resources as well as biodiversity.[7] Finally, ideal food systems are **resilient**. They must be able to bounce back quickly from more frequent health, climate, and economic shocks, and also provide poor households with stable livelihoods that protect them from these shocks. Strengthening these five food system attributes requires an enabling environment for optimal food system policies, governance, and accountability (Chapter 2).[8]

COVID-19 has highlighted the risk of shocks to our food systems that can lead to multiple supply and demand disruptions. At the same time, ongoing climate-change-induced weather shocks continue. Beyond these, other shocks affected poor countries in 2020. Falling oil prices contributed to declines in incomes and food security in oil-exporting developing countries. Torrential rainfall in East Africa triggered swarms of locusts, and many countries experienced destructive flooding, typhoons, wildfires, hurricanes, and tropical storms that broke natural disaster records. With the likelihood that shocks, including pandemics and natural disasters, will only become more frequent (Chapter 4), resilience from the farm to the global level must be a focus for food system transformation.

COVID-19 IMPACTS ON FOOD SYSTEMS

The pandemic and the policy responses adopted to address it have affected our food systems from the global to the local level, setting back already-uneven progress and exposing weaknesses and vulnerabilities. In IFPRI's recent book *COVID-19 & Global Food Security*, we identified many different impacts of the pandemic and pandemic responses on the world's food systems. This report expands on those insights, looking at what we have learned in this first year of the pandemic, with a view to transforming food systems for the long term.[9]

LOST INCOME

The pandemic's impacts on food security have been induced primarily by falling incomes. The World Bank estimates that the global economy shrank by 5 percent in 2020, with the greatest burden borne by poor people. By the end of 2020, 95 million additional people, mostly in Africa south of the Sahara, were estimated to be living in extreme poverty.[10] IFPRI research estimates that the number of poor people globally is likely to increase by about 150 million, 20 percent above pre-pandemic poverty levels.[11] The experiences of many countries confirm the importance of lost incomes. A survey of mothers in Myanmar found that median incomes declined by a third, leading to a 27 percentage-point increase in income-based poverty over six months. In China, 18 percent of SMEs had closed permanently by May 2020 (often due to lack of consumer demand), representing a loss of 14 percent of total jobs.[12]

Remittance income was particularly affected because of the pandemic restrictions on travel and movement. In 2019, remittances represented more than 10 percent of gross domestic product (GDP) in seven African countries, and were also important for low-income Asian and Latin American countries.[13] As a result of the pandemic, the global flow of remittances has fallen by almost 20 percent, and flows to Africa by 23 percent. This report's Regional Developments section examines this challenge in many countries, from Yemen, where the reduction in remittances lowered household incomes by 12.5 percent, to China, where more than 10 percent of low-income remittance-receiving rural households are expected to fall back into poverty. Recent research also suggests that restrictions on global travel and freight put into place by rich countries to stop the virus's spread represented a larger economic cost for poor countries than their own pandemic restrictions.[14]

FIGURE 1 Food system transformation goals

Source: Based on S. Fan et al., "Food Systems for Human and Planetary Health: Economic Perspectives and Challenges," *Food System Economics* (forthcoming).

DISRUPTION OF FOOD SUPPLY CHAINS

Food supply chains were disrupted by labor restrictions and falling demand, although impacts varied along the value chains and between countries and commodities (Chapter 6). Food services were particularly affected, and many poor people lost jobs in urban

Source: IMF, *World Economic Outlook*, January 2021 Update (Washington, DC: 2021).

TIMELINE OF Global COVID-19 Outbreak

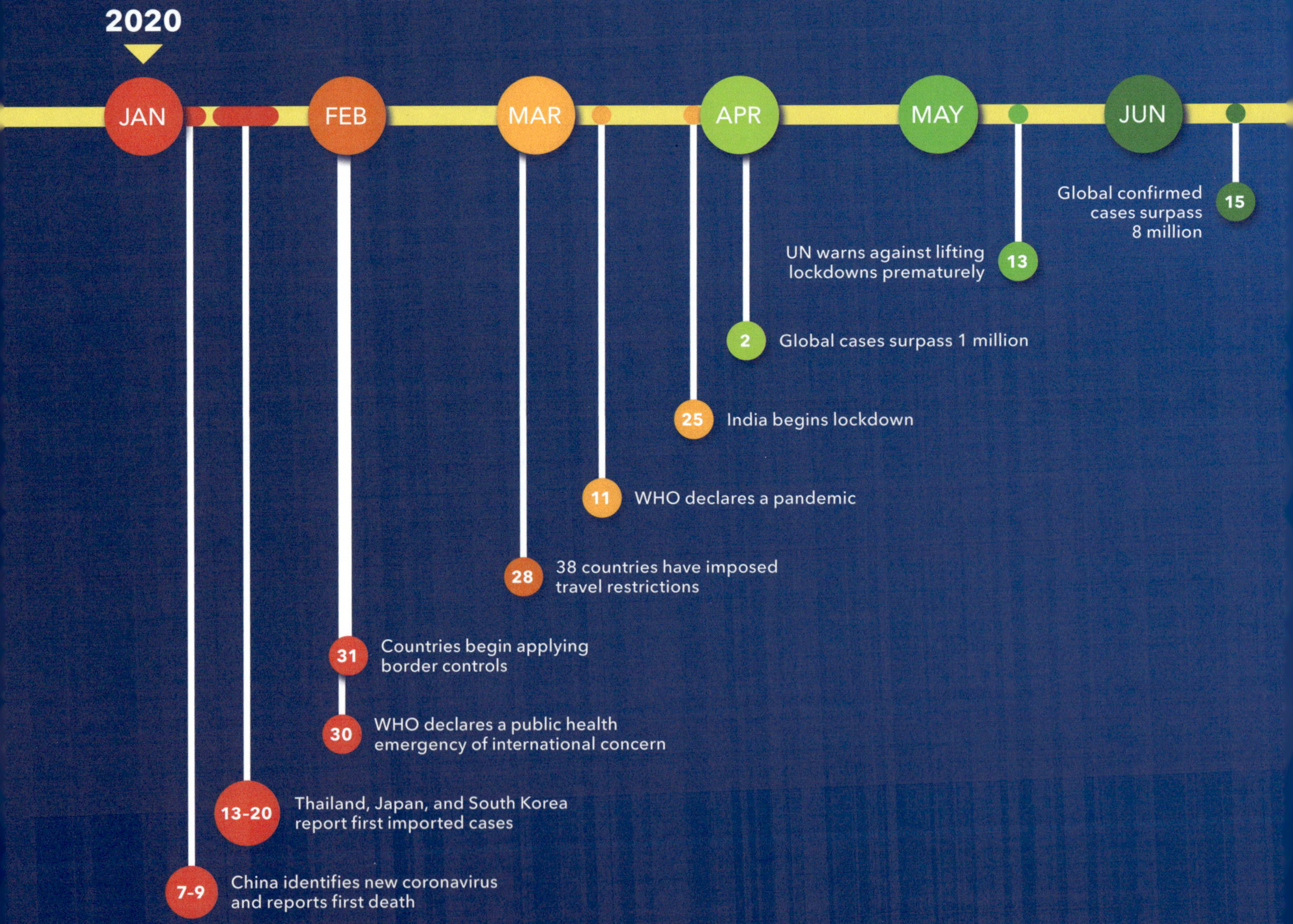

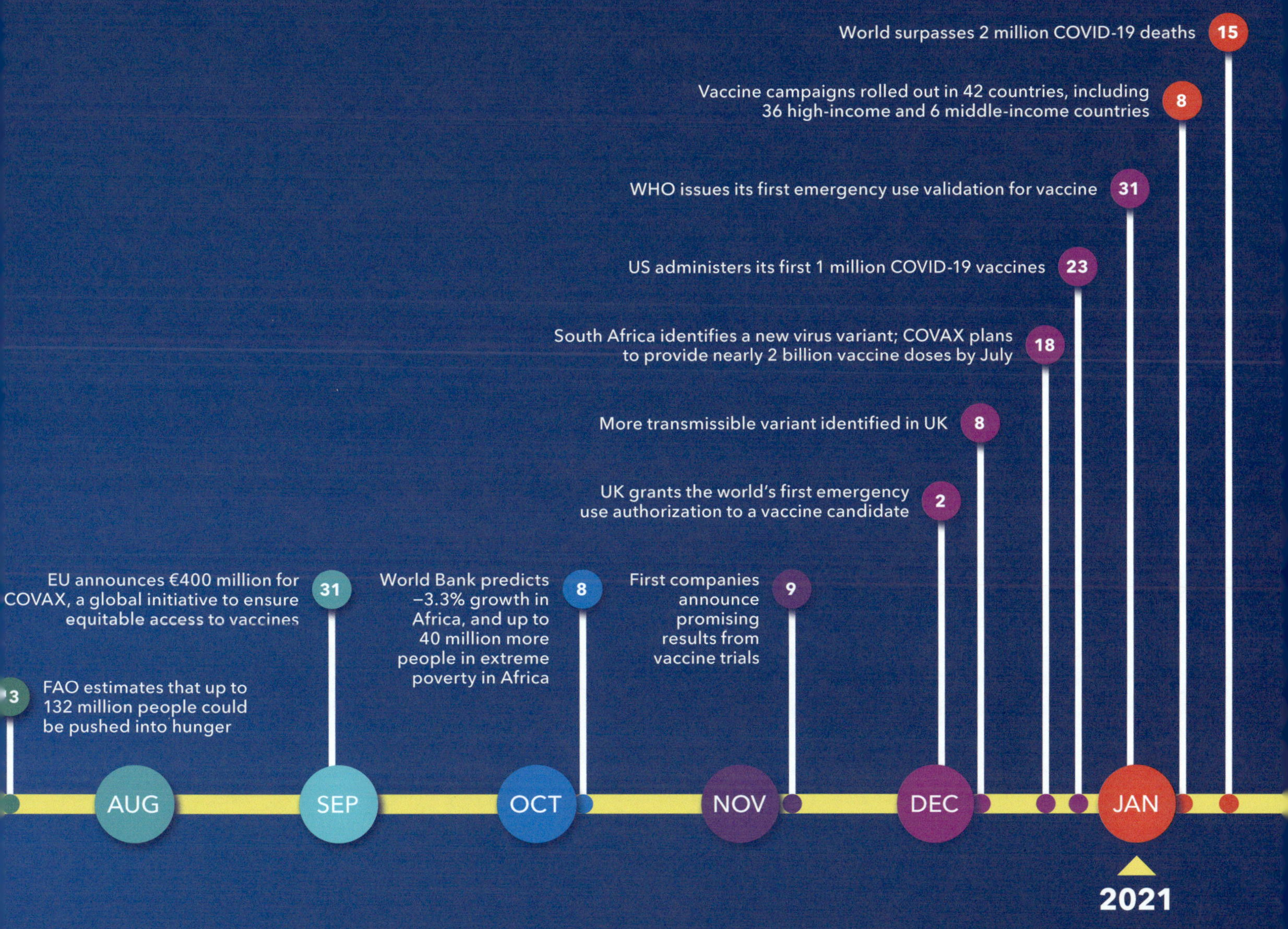

CUMULATIVE CONFIRMED COVID-19 DEATHS

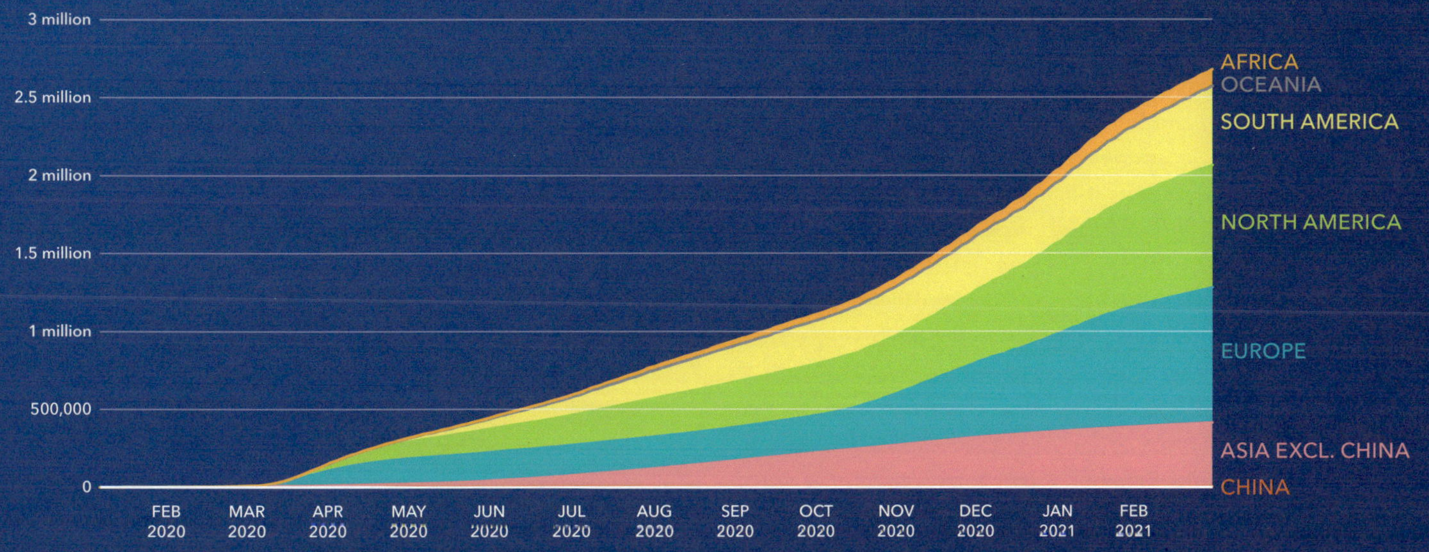

Source: Johns Hopkins University, CSSE COVID-19 Database (updated March 18, 2021).

areas, especially in the tourism and restaurant sectors. Traditional food systems, with few linkages beyond the farm, and modern, vertically integrated systems were relatively resilient. More vulnerable, however, were food systems transitioning from traditional to modern, which are characterized by longer supply chains and still-fragmented storage, transportation, and services. During the pandemic, these characteristics made it more difficult for actors along transitioning food supply chains to access markets and to procure labor and input supplies, which were affected by drops in demand and government restrictions. For example, in Myanmar, half of retailers of agricultural machinery, such as tractors and combine harvesters, reported a drop in sales of 70 percent or more. They attributed this to movement restrictions as well as decreased machinery imports and decreased local production of machine parts.[15] Modern, vertically integrated supply chains, with more options in terms of suppliers and logistics as well as the ability to move many transactions online, were most able to adjust to pandemic conditions.

Impacts on agricultural production and trade were mixed. Both labor supply and perishable products were notably affected by mobility restrictions. In Senegal, for example, small fruit and vegetable producers were impeded by closure of traditional wet markets and social-distancing requirements, leading to food spoilage and lost income.[16] In many countries, the pandemic struck during the agricultural season, and mobility restrictions reduced the availability of labor for harvesting and other farming activities, with the most severe impacts on farms that rely heavily on hired labor. In Ethiopia, for example, restrictions on mobility constrained the labor supply to rice farmers, about 75 percent of whom rely on hired day-laborers for weeding and harvesting.[17]

Some countries responded to the pandemic with trade restrictions reminiscent of their policy reactions to the 2008/2009 food price crisis. Restrictions on exports can reduce incomes of producers, and run the risk of triggering food price spikes in importing countries.[18] According to IFPRI's Food Trade Policy Tracker, 19 countries introduced export restrictions, with severe effects on importing countries, including some of the poorest countries in Central Asia and Africa south of the Sahara.[19] Kazakhstan's ban on exports of wheat and other products in March 2020 affected 50 percent of neighboring Kyrgyzstan's food imports.[20] However, many of these restrictions were removed or loosened in the second half of 2020, following strong reactions from international organizations.

FOOD INSECURITY AND NUTRITION LOSSES

The impacts of rising poverty and reduced livelihoods are reflected clearly in rising levels of food insecurity and decreasing diet quality. For example, more than a third of Bangladeshi rural and urban youth reported moderate or severe food insecurity during the country's lockdown period. Unsurprisingly, the highest prevalence of food insecurity was found in groups that had reported losing household income.[21] Coping mechanisms reported by poor Bangladeshis included not eating for an entire day and exhausting household food reserves. Rural Nepalese households found themselves in a similar situation: preliminary surveys found that even six months after an initial lockdown, 40 percent of households were using their savings to cope, and more than 30 percent had reduced their spending on food items.[22]

Dietary quality and diversity were affected foremost through declining incomes but also through trade and movement restrictions that have disproportionately affected the availability of nutritious perishable products. Several studies, including from Guatemala and China, illustrate households' dietary shift away from more expensive nutritious foods, such as fruits, vegetables, and animal-sourced foods, toward cheaper staple foods.[23] In addition, lockdowns affected food security and nutrition through public sector channels. Lockdowns shuttered schools and daycare centers, which provide critical meals and supplementary nutrition to hundreds of millions of young children. India's Mid-Day Meal program, for example, covers 80 percent of primary-school-aged children in the country, improving not only nutrition but also learning outcomes and gender equity; the country's school closures likely exacerbated food insecurity and malnutrition, especially for girls and disadvantaged populations.[24]

The decline in food security and sound nutrition will have long-lasting development effects. Early in the pandemic, IFPRI researchers and their colleagues projected that even short lockdown measures implemented in tandem with restrictions on mobility and food system disruptions would lead to a 7 to 9 percent decrease in gross national income (GNI) in most low- and middle-income countries (LMICs), as compared with pre-pandemic projections. Applying these estimates to 118 LMICs suggested that moderate or severe wasting (low weight-for-height) among young children could increase by 14.3 percent, adding an estimated 6.7 million more wasted children in 2020.[25]

MAGNIFYING DIFFERENCES AND SHORTCOMINGS

The pandemic has been a stark reminder of the greater vulnerability of the **poor** and other disadvantaged groups. Wealthier households in LMICs generally experienced larger percentage declines in income, primarily because they were likely to work in industry and service jobs that were disrupted by COVID-19 shocks and restrictions. Poor households, however, suffered far more detrimental impacts on food security, livelihoods, and wellbeing. Because these households spend a larger share of income on food, so loss of income increases their food insecurity, and they have fewer assets to help them mitigate shocks.[26] In Ethiopia, for instance, poorer households were far more likely to report high stress than were wealthier households.[27] Food security and livelihoods were most negatively impacted in places where food value chains were poorly integrated.[28] The pandemic has also disrupted vital services upon which poor communities rely, including public sector programs for poverty alleviation, health, and food and nutrition, such as school-feeding programs and work-for-food programs.

The impact of COVID-19 on the **rural-urban** gap is more complex.[29] On the whole, urban households experienced larger income losses because they rely on income from industry and service sectors that was severely affected by lockdowns and economic recession. For many urban poor, their sole productive asset is their physical labor, which many were prevented from using due to pandemic-related restrictions. Agriculture and agrifood systems, with the exception of food services and restaurants, have been more resilient, both because consumers prioritize food when incomes decline and because agrifood systems were generally exempted from lockdowns and movement restrictions. For this reason, the rural poor who depend on income from agriculture and other agrifood activities were partially buffered from the shock. Nevertheless, poverty has risen in both urban and rural areas, and due to larger rural populations living close to the poverty line, rural areas account for more of the people pushed into poverty. This reflects the particular vulnerability of rural households and underscores how even small income losses can lead to critical deteriorations in welfare. Falling remittances from urban workers and workers abroad have also spread the effects of urban income losses to the rural poor.

Disadvantaged groups have suffered most during the pandemic, due to the economic, legal, and social barriers they already faced and their reliance on informal work. **Women** account for 39 percent of employment globally but incurred 54 percent of total job losses during the pandemic, reflecting their heavy representation in informal activities.[30] In many poor countries, women have experienced increases in their already-heavy workloads. In an April 2020 survey conducted in rural India, about 50 percent of households reported that women were spending more time fetching water and firewood, in comparison with earlier in the year.[31] Stress on households sparked by lost incomes and stay-at-home orders too frequently leads to increased domestic violence that most affects women and children. In Peru, researchers reported that calls to the national domestic violence hotline increased by 48 percent between April and July 2020.[32] Yet, national policy responses have largely failed to adopt a gender-sensitive approach, and risk leaving women further behind. For example, Myanmar increased the availability of low-interest loans to farmers, an opportunity that women are less likely to make use of than men because women are rarely legal landowners.[33]

Responses to the pandemic also amplified the **digital divide** between rich and poor. Rich communities were able to rely on Internet services to access schooling, market information, health services, and more, while poor communities were left in relative isolation. These impacts may be long-lasting; for example, disruptions in schooling will likely lead to lower lifetime earnings, poorer health, and less opportunity for many children in developing countries to escape poverty.

Beyond this broad-brush description of the gaps exposed by the pandemic, the Regional Development section in this report also identifies important **regional and national differences** in policy reactions, demographics, food and economic system structures, and how these influence the pandemic's impacts on citizens. How the pandemic will evolve in different regions and its long-term impacts are still unknown; we have just a snapshot from 2020.

Countries in South Asia, for example, initiated strict lockdown measures at the earliest stage of the pandemic, keeping COVID-19 cases low, and spent sizable resources on their already-large social protection networks, on support to agriculture, and on maintaining food price stability. But the region's labor markets, especially in the nonfarm and informal sectors, proved to be fragile, due in part to their reliance on migrant workers. The drastic reduction in international remittances was also a big blow.

Similarly, countries in East and Southeast Asia suffered from the loss of international and domestic remittances, with rural households suffering the most. The region has also faced a challenge in reaching the urban poor, many of whom work in the informal sector and are not covered by existing social protection systems. Expansion of regional trade, however, has helped East Asian countries mitigate the impacts of global trade restrictions; many exports, including agricultural products, have been rerouted to China.

Central Asian countries quickly enacted policy measures to contain the virus, but still experienced a contraction in GDP, closure of SMEs, and isolation of some segments of society, especially in rural areas with poor digital connectivity. The pandemic also exposed the weaknesses of many of the region's economies, including dependence on remittances and on a few trading partners and commodities subject to price volatility. Fortunately, agricultural growth remained relatively robust and intraregional trade mitigated some of the pandemic's effects.

Africa south of the Sahara has recorded relatively few COVID-19 deaths, but short-term policy responses and the global recession have interrupted 25 years of economic growth, with decreases in household incomes, lost employment, increased poverty and food insecurity, and protests over lockdowns. Some countries were able to invest heavily in social protection measures, but rates of coverage remain low compared to the global average, and the high costs of pandemic-response programs may risk a fiscal crisis.

Latin America and the Caribbean have been hard hit, due to a high level of urbanization and the ease with which the virus spreads in dense environments. As in other developing regions, employment in the informal sector, which lacks social safety nets, is common and has been severely disrupted by movement restrictions. The region also has among the highest rates of obesity and overweight, increasing the death rate, and for many, the severity of the disease.

Finally, in the Middle East and North Africa region, the pandemic led to falling remittances and incomes, especially in the service and industry sectors. Food services and tourism-related businesses suffered the most severe disruptions, disproportionately harming urban dwellers employed in those sectors, while other parts of the agrifood system have proved more resilient.

A UNIQUE OPPORTUNITY FOR TRANSFORMATION

Despite the many negative impacts of the pandemic, the health, economic, and food disruptions have opened opportunities for fundamental change. COVID-19 magnified many long-term weaknesses, such as persistent inequalities and poorly integrated supply chains, that must be addressed. But it also highlighted the benefits of investments and policies that have created an enabling environment for private sector innovation, with flexible markets and appropriate infrastructure, and of sound policy systems that are able to respond quickly and nimbly. The chapters in this report examine the lessons learned about what worked well and can provide building blocks for food system transformation.

As of early 2021, there is considerable hope that the rapid development and deployment of effective vaccines will relieve the disease threat. However, the emergence of new coronavirus variants, the difficulty of ensuring an adequate supply of vaccines, and the challenge of vaccinating all people, particularly the poor, is almost certain to prolong recovery in many places. Although it seems increasingly likely that the world will have to learn to live with the virus, its impact will ebb,

BOX 1 INTERNATIONAL EVENTS TO PROMOTE FOOD SYSTEM TRANSFORMATION

Several major international policy events planned for 2021 can help build the necessary political will and provide momentum for overhauling food systems to meet global goals. These global initiatives will help countries to tap into funding, research, and communities of practice to build capacity for change.

The 26th United Nations Climate Change Summit (COP26) may be the most important UN summit since the Paris Agreement. Amid worrying reports on the world's GhG emissions trajectory, despite the dip in emissions during the pandemic, COP26 represents a chance for countries to pivot their post-pandemic recovery plans toward environmental sustainability. Given the large contribution of plant and animal agriculture to GhG emissions and other natural resource degradation, transforming food systems to be sustainable is critical to achieving environmental and climate change goals.

The UN Food Systems Summit aims to "launch bold new actions" to make progress on the SDGs, explicitly recognizing that achieving each of the goals will require sustainable, healthier, and more inclusive food systems. The event is a powerful call for action to change food systems at all levels. It also aims to create a system of accountability, under which countries will monitor and report progress toward food system transformation.

The Tokyo Nutrition for Growth Summit will encourage governments, businesses, multilateral organizations, and donors to make concrete financial, programmatic, and impact commitments in three focus areas: health, resilience, and transforming food systems so that they promote safe, sustainable, and healthy foods. Other events in 2021 where food system transformation should be considered include the World Trade Organization Ministerial Conference and the UN Biodiversity Conference (COP15).

allowing attention to focus on the longer-term agenda of transforming food systems to be healthier and more efficient, sustainable, inclusive, and resilient.

COVID-19, like other crises, has triggered reactions from governments, the private sector, farmers, consumers, and the international development community, many of whom altered their roles, operations, and behaviors in ways that were previously constrained by a variety of political, social, technical, and economic barriers. This normalization of out-of-the-box approaches has fundamentally changed thinking about the potential of food system transformation, making this the right time for the deep changes that are needed. As the Global Panel on Agriculture and Food Systems for Nutrition's latest Foresight report states, transformation will depend on "the political will and courage to reform outdated policies and a sustained commitment to act."[34]

Upcoming global summits and new thinking have the potential to be catalytic (Box 1), but the real transformation must occur in regions, countries, and communities through policies, investments, and actions that adapt and build on past successes and address weaknesses. Financing these changes will also require innovative approaches and mechanisms from the global to the local level to support public and private sector investments in transformation (see the special section: Financing the Transformation to Healthy, Sustainable, and Equitable Food Systems, following this chapter). What is encouraging is that the pandemic has already triggered transformations within the public and private sectors. In many cases, these short-run actions meant to deal with an immediate crisis will have long-run benefits.

Since the pandemic's onset, governments have adopted a variety of response policies, from increasing spending on health systems and vastly expanding social protection to supporting private businesses.[35] Social protection efforts served to test the effectiveness of pro-poor interventions and policies. Ethiopia's flagship Productive Safety Net Program, for example, offset nearly all of the pandemic's negative impacts on the food security of participating households, especially poorer households and those living in remote areas[36] In India, efforts were made to incorporate migrant workers into social protection programs;[37] and in Bangladesh, cash transfers were substantially expanded. Other efforts focused on the informal sector, composed of small vendors who were especially hard hit: Burkina

Faso, for example, created a fund for women who sell fruits and vegetable in informal markets. Such actions showcase ways to support small actors who are critical to urban food systems.[38]

The private sector's experience sheds further light on how food systems can become more resilient. Some food systems have proved to be more resilient than others, depending on their structure, ability to quickly adapt to shocks, and the government's role in supporting value chains.[39] For example, large, vertically integrated fruit and vegetable companies in Senegal, which can control their suppliers, distributors, and retail locations, were able to provide protective gear and safer transport and invest in cooled storage capacity; as a result, they fared better than smaller, less integrated companies.[40]

Almost everywhere, businesses that were able to digitalize quickly, from food delivery in urban areas to market information provision in a mobile format, proved to be far more resilient throughout the pandemic. Their experience offers lessons about the potential for the digital revolution – from drones monitoring crop quality to the urban poor accessing mobile banking – to make food systems more resilient.

Some innovations foster resilience while also contributing to other attributes of an ideal food system. For example, the emergence of SMEs devoted to solar-powered transportation and cold storage of fruits and vegetables can help make local food systems more resilient, environmentally sustainable, nutrition-driven, and inclusive.

BUILDING A MORE RESILIENT FUTURE

The innovations in the public and private sector in response to COVID-19 are encouraging, but much more remains to be done. The chapters in this report provide a strong set of policy recommendations targeted to different aspects of food system transformation and to different regions and countries. Many of these focus on inclusion, efficiency, health, and sustainability. The pandemic is above all a test of resilience to a shock, and so in this final section we focus on the lessons about resilience in food system transformation. For many countries, the end of the pandemic is not yet in sight, and other shocks – including new diseases, conflicts, natural disasters, and climate-change-induced disruptions – are likely to become more frequent.

Making food systems more resilient requires a set of actions, many of which must be adjusted to local circumstances and food system characteristics. Three types of measures are needed. First, the best way to build resilience into our food systems is through shock prevention, and when shocks can't be avoided, to **limit the frequency and magnitude of shocks**. Some ongoing vulnerabilities, such as those due to climate change and inequality, were exacerbated during the pandemic. Therefore, investing in mitigating predicted multiple shocks now – for example, investing in climate change mitigation – will reduce the likelihood and magnitude of various shocks, such as droughts and flooding, in the future. Another example is reducing inequality. Promoting inclusiveness in economic systems is likely to reduce or prevent social conflicts that are an important source of food insecurity and welfare declines.

Second, resilience implies the capacity to **anticipate shocks**. Information is crucial to help people, businesses, and governments prepare for shocks. Investments in early warning systems, development of improved data and indicators, and digital technology are examples of ways to increase access to information. In the case of COVID-19, dynamic metrics for tracking the transmission of the virus, including speed, acceleration, and persistence of COVID-19 cases, and indicators of the impacts of policy responses are useful. Similar indicators should be developed for tracking other potential shocks such as climate events, civil conflict, and pest infestations.[41] The Food Security Portal, facilitated by IFPRI and supported by the European Commission, provides data on dynamic developments in food systems around the world, including food price volatility, so that policymakers can respond in a timely way.[42] (See more about the resources offered by IFPRI on page 18).

Increasing access to information and communications for everyone can play a vital role in building capacity of diverse groups to strengthen resilience. The rise of digital tools and services during the pandemic illustrated how unequal access can affect people's lives and livelihoods. Governments must aim to close the digital divide and ensure that all their citizens have access to the benefits of the digital revolution, particularly in food systems, by investing in rural energy, mobile ICT networks, and big data analytic systems geared toward smallholders and disadvantaged communities. The private sector also has a critical role to play, as its investments in digital services and e-commerce can open opportunities for the integration of small farmers, SMEs, and consumers in future food systems.

Third, improving the capacity of all actors in our food systems to **absorb shocks** is the final piece of the resilience puzzle. Capacity enhancement requires a variety of instruments, such as better access to finance (liquidity); flexible social safety nets; lower transaction costs in value chains; competitive markets for inputs, outputs, and logistics; reliable trade agreements; investment in rural services, infrastructure (including digital connections), and R&D for improving food production systems; and more. At the global and national levels, multilateral financial institutions will need to address the liquidity constraints of many developing countries. Small producers and SMEs need access to credit, capital, and insurance to mitigate risk. Social safety nets can protect the most vulnerable people from shocks and also lead to gains in welfare and food and nutrition security. Conditional cash transfer programs, for example, have proven impacts on poverty reduction, household food consumption, and dietary diversity.[43] These transfer programs can also be used to build up women's control over resources, enhance their empowerment, and strengthen their social networks.

In sum, a wide set of measures are needed to make our food systems more resilient. The ongoing pandemic has shone a harsh light on the vulnerabilities of our food systems, but has also proved that food systems can be resilient and that adaptations and innovations can be greatly accelerated. Food systems in developing countries have typically been less resilient and more vulnerable, causing the greatest harm to the poor and disadvantaged. Looking forward, measures for resilience need to be embedded in longer-term transformation strategies to make food systems more efficient, inclusive, sustainable, and healthy. In addressing resilience, we must pay special attention to the most vulnerable households and communities in our food systems.

All this requires a purposeful transformation of our food systems, globally and locally. Careful research and analysis are required to identify the most effective measures for such a transformation. The chapters in this report provide a series of evidence-based ideas and recommendations – supported by high-quality research, some produced over a span of decades and some in the midst of the pandemic – for making such a transformation possible.

TOOLS AND RESOURCES

TOOLS

AGRICULTURAL PRODUCTION AND STOCKS MONITOR

Visualizing production and stocks of key crops at global and country levels, with comparisons to levels during the 2008–2009 food price crisis.

COVID-19 FOOD PRICE MONITOR

Providing daily updates of food price movements in wholesale and retail markets in key countries in South Asia and Africa south of the Sahara.

COVID-19 POLICY RESPONSE PORTAL

Capturing policy responses to the pandemic, including population restrictions, social protection, trade, health, fiscal, and monetary measures.

FOOD TRADE POLICY TRACKER

Monitoring restrictions on food exports and trade and their impacts on food imports.

RESOURCES

MEASURING IMPACTS AND PRIORITIZING POLICIES FOR RECOVERY

EXPLORE

IFPRI's researchers, working together with governments and local partners in many African and Asian countries, are evaluating the pandemic's economic costs and identifying policy priorities for relief and recovery.

COVID-19 BLOG SERIES

EXPLORE

This special series of IFPRI blog posts, running since the onset of the pandemic, analyzes the impacts on national and global food and nutrition security, poverty, and development.

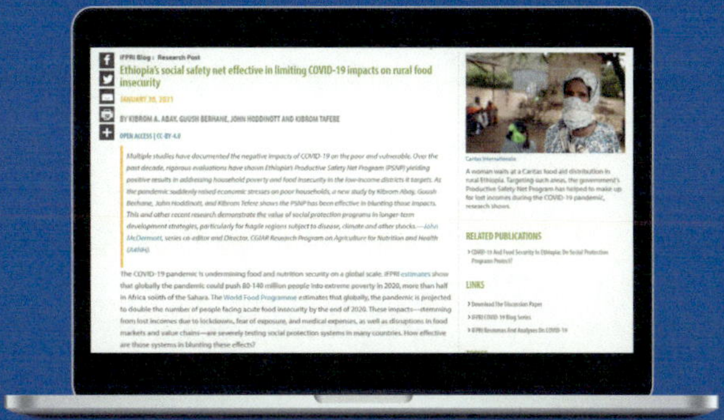

To learn more about the impacts of COVID-19 around the world, visit our webpage on "IFPRI Resources and Analyses of COVID-19 Impact" for events, blogs, videos, podcasts, and publications, as well as access to additional tools and resources:

ifpri.org/covid-19.

RESEARCH HUB

CGIAR COVID-19 HUB

EXPLORE

The COVID-19 Hub provides a coordinated research response to the global pandemic, focused on national response and recovery work across CGIAR research themes. Convening researchers, funders, and key stakeholders, the Hub works across four research areas to support country responses and address food system fragilities to build back better. Hosted by the CGIAR Research Program on Agriculture for Nutrition and Health.

DATASETS

DATA FOR FOOD SYSTEMS RESEARCH

EXPLORE

By providing open-access datasets, IFPRI supports research, innovation, and evidence-based policymaking. More than 500 primary and secondary datasets cover a range of topics relevant to food systems drawn from experiments, field trials, simulations, geospatial analysis, macroeconomic analysis, and socioeconomic household and community-level surveys. Datasets are shared through the IFPRI Dataverse online repository, hosted by Harvard Dataverse.

Financing the Transformation to Healthy, Sustainable, and Equitable Food Systems

EUGENIO DÍAZ-BONILLA, JOHAN SWINNEN, AND ROB VOS

Eugenio Díaz-Bonilla is head of the Latin America and Caribbean Program, International Food Policy Research Institute (IFPRI), Washington, DC. **Johan Swinnen** is director general of IFPRI, Washington, DC. **Rob Vos** is director of the Markets, Trade, and Institutions Division, IFPRI, Washington, DC.

Our agrifood systems must undergo profound change to address the interlocking challenges of persistent rural poverty, climate change and environmental degradation, and the triple burden of malnutrition. Food systems are largely market driven – through the production choices of farmers and large and small businesses in food trade, transport, distribution, and services, as well as by consumer choices. But governments can steer the decisions made by these actors through policies, regulations, and investments in public goods (such as infrastructure), many of which are discussed in this report. These instruments can also leverage private sector finance for investments in food systems. Transformative change toward climate-resilient, sustainable, inclusive, and healthy food systems, however, will require fundamental reorientation of market incentives and investments.

Purposeful transformative change will also require adequate finance. Estimates of the financing needs for achieving the UN's Sustainable Development Goals (SDGs) are substantial, ranging from US$50 billion to well over $1 trillion per year over the next decade, depending in part on which SDGs are included.[1]

How can we mobilize the necessary resources? Here, we present a basic framework of key fiscal and financial mechanisms to help orient an actionable agenda. This framework is built around six intervention areas for both incentive reform and financial resource mobilization to promote investments in food system transformation: (1) consumer expenditures on food; (2) agrifood business profits and savings; (3) fiscal measures (public expenditures and taxes); (4) international public finance (official development assistance [ODA] and nonconcessional lending by bilateral donors and multilateral development banks [MDBs]); (5) bank finance; and (6) capital market finance (Table 1). The first two areas of intervention and financing sources can be considered "internal" to food systems' economic actors, while the other four are "external" to them.

Food system transformation will require a fundamental rethinking of existing financing mechanisms. This rethinking should be guided by four key objectives: (1) ***repurposing*** finance and public support to promote better food systems in each area/funding source; (2) ***reducing*** finance that is destroying or degrading the natural-resource base of food systems; (3) ***optimizing*** finance through de-risking investments in better food systems; and (4) ***scaling up*** and leveraging finance flowing to investments in better food systems.[2]

Consumer expenditures are a critical driver of food system outcomes. They not only reflect dietary choices and people's ability to access food, which are key to nutritional and health outcomes, but they also form the demand side of the market for food producers and intermediaries. Consumer spending determines the revenue stream for food retailers and for suppliers in both downstream and upstream segments of food supply chains, thus influencing potential net profits of these actors and their investment choices. Therefore, public policies that affect consumers' incomes and their food options and choices will also shape investment decisions and financing requirements of farmers, agrifood businesses, and commercial and food service operators. Policy tools that can be useful for influencing food demand (both its level and composition) include nutritional and environmental education, targeted taxes and subsidies, and food standards and other regulations, as well as income and nutrition support through social safety nets.

Significant **private sector investments** are needed to transform our food systems and achieve the SDGs. These efforts will have to come from all private actors – from small-scale farmers and agrifood businesses to large-scale global and national food processors, retailers, and aggregators. Public policies can influence the size and allocation of those private investments and operating expenditures. Macroeconomic and trade policies, for instance, can improve the general business environment by fostering price stability and competitiveness for domestic production. Through targeted provisioning of public goods, such as support for appropriate education and skills training, R&D, and basic infrastructure, governments can promote a wide range of changes, such as better access to markets for smallholders and the development and adoption of sustainable production. Within food value chains, institutional reforms and policy initiatives that help improve market functioning and de-risk private sector investments will attract resources for investments that would otherwise not occur. Such initiatives would typically aim to improve the competitive operation of markets, strengthen contract-farming systems, improve access to value chain finance, and strengthen risk management. Credit guarantees and robust markets for innovative insurance, such as parametric (for example, weather-indexed) insurance, could help strengthen risk management. Environmental regulations (such as those protecting forests and ecosystems) plus expanded programs of payment for environmental services, along with food quality regulations and standards and stricter labeling requirements, could trigger shifts in private investment toward sustainable production of more nutritious foods.

Fiscal measures intended to influence consumer demand and agrifood sector profits and investments for food system transformation will have to be aligned with overall government fiscal priorities, on both the expenditure and revenue sides. In the context of a broader public expenditure review, an essential starting point would be to rethink existing subsidies and other support measures for agricultural production and food consumption. Based on OECD estimates for 54 countries, these support measures added up to about $600 billion per year in 2017–2019, of which about 56 percent was incurred as a direct fiscal cost in the form of coupled and uncoupled subsidies and other general support expenditures for producers; 11 percent as fiscal expenditures that support food demand by consumers; and the remainder being the imputed costs of international trade measures, primarily implicit transfers from consumers to producers.[3] These agricultural support measures have been maintained over a long period of time in many countries and have contributed to the current unhealthy and environmentally unsustainable production and consumption patterns. Repurposing this existing agricultural support to forms that align with the objectives of food system transformation thus seems to offer a first important step toward reorienting both consumer demand and private investment decisions. It should be noted that currently more than half of all agricultural support is provided by developing countries, though only abut $50 billion of that is provided in the form of subsidies by developing countries other than China.

However, countries with limited fiscal resources have little funding that can be repurposed. The global recession provoked by the pandemic and lockdown measures have severely weakened the fiscal position of most countries in the world, particularly low-income countries, limiting their capacity to counteract adverse economic impacts and protect the poor.[4] Economic recovery is likely to be slow in low- and lower-middle-income countries and may be quite challenging without sustained external financial support to strengthen their budgets.[5]

Here, concessional and nonconcessional **international public finance** could make a difference, if scaled up substantially. According to the latest numbers (2019), current funding allocated to agriculture, rural development, and food security is limited, amounting to about $17 billion per year, of which about $12 billion is provided as ODA, with an additional $2 billion for emergency food assistance.[6] The global recession has put further pressure on already vulnerable donor aid budgets. Nonconcessional loans from the World Bank and regional MDBs mainly go to middle-income countries, and net flows (that is, disbursements minus repayment of previous loans) are even lower than those of ODA. Substantially increasing net flows would require commensurate expansion of the capital base of the MDBs. Yet, even if somewhat limited in scale, lending from these sources could have greater impact if used strategically to support the realignment of national public spending mentioned above and to leverage blended forms of finance designed to de-risk private sector investments in sustainable food systems.

External sources of financing – namely, loans from the banking system or financial operations in capital markets – can expand and supplement the internal sources mentioned. Financial institutions and investors tend to perceive investments in agriculture and agrifood systems as being of high risk and low return. Regarding the **banking system**, existing macroeconomic policies, regulatory frameworks, financial instruments, and delivery systems need to be carefully reviewed in each country context to identify the obstacles that constrain access to finance for agrifood businesses, especially for small-scale farms and small and medium enterprises (SMEs). Removing or reducing these obstacles within the broader financial system is important to ensure access for these and other often-marginalized actors. Many efforts to improve financial inclusion have focused on specialized mechanisms, such as microcredit schemes or mobile payment systems for the poor, but although these mechanisms can improve financial inclusion, they generally cannot provide finance on the scale needed to induce transformative change. For this, credit supplies from the broader banking system will be needed, as well as other financial services for farmers and SMEs in food value chains.

To this end, improving the business environment and incentives for agrifood system transformation, including the instruments for de-risking as well as fiscal and macroeconomic measures that raise the profitability of sustainable and healthy food production will be essential. Depending on the context, strategic use of government or donor-supported development finance to mobilize commercial bank lending, as well as provision of seed money for the development of cooperative savings and loans associations and digital payment platforms for farmers and agrifood businesses could catalyze significant shifts in commercial bank lending. As part of this reorientation of financial systems, renewed consideration could be given to allowing national development banks as well as central banks to engage in development financing in low- and lower-middle-income countries with underdeveloped capital markets. In the 1960s and 1970s, both types of financial institutions played a substantial role in financing rural and agrifood sector development in a number of developing countries through dedicated lines of credit. In subsequent decades, the role of national development banks and "developmental central banks" was largely curtailed, a decision that should be reassessed.[7]

Global and national **capital markets** offer another major potential funding source for transformative investments in agrifood systems. For instance, in 2020, the issuance of "green bonds" exceeded US$1 trillion. Other options are impact investors, investors with environmental, social, and governance (ESG) requirements, sovereign funds, and a variety of other potential institutional and private investors, all provided that the investable vehicles and financial structures meet the risk/reward profiles and other conditions that those economic actors expect. At present, however, investments in projects and financial vehicles related to agrifood system transformation are small, possibly because many ventures are considered high risk. Also, transaction costs to develop "bankable" projects and to design and launch new categories of asset classes are high. To overcome these hurdles, one possibility is to create a special *facility for project preparation, incubation, and acceleration*. This facility would help investors and food sector actors develop a portfolio of projects and investable financial instruments with the adequate balance of risk and return to mobilize the funds of investors seeking stable, long-term returns while supporting aspects of the SDGs, such as socially and environmentally sustainable food production by small producers.[8]

Additionally, as noted, official international finance, public sector expenditures, and even philanthropic funds can be used as blended finance, guarantees, absorption of "first losses," and other means to de-risk private investments. Digital innovations (such as crowdsourcing and investment-opportunity exchanges) could also reduce the transaction costs of connecting small and medium investors with options for financing transformative activities in food systems.

In sum, substantial financial resources will be needed to transform food systems in ways that contribute to achieving the SDGs. The variety of funding sources and policy interventions to finance such transformation outlined here should be further explored and acted on. The first-ever UN Food Systems Summit in 2021 will provide a crucial forum for transforming these ideas into concrete action areas to *repurpose, reduce, optimize, and scale up* finance, and to identify how financial innovation can contribute to the transformation of our food systems.

TABLE 1 Sources of funding and areas for realignment of investment incentives

AREAS FOR INCENTIVE REFORM AND FINANCING SOURCES	POSSIBLE POLICY INSTRUMENTS
CONSUMER SPENDING ON FOOD	▪ Nutrition education ▪ Regulations and stricter labeling requirements and food standards to support healthy, sustainable diets ▪ Consumer taxes on unhealthy and unsustainably produced foods and subsidies to support healthy, sustainable diets ▪ Social safety nets with nutritional aspects
FARM AND AGRIFOOD BUSINESS PROFITS AND SAVINGS	▪ Macroeconomic and trade policies for sustainable, healthy diets ▪ Producer taxes on unhealthy and unsustainably produced foods, and subsidies to support nutritious and sustainable food production ▪ Provision of public goods (infrastructure, R&D) ▪ Policies and regulations for market competition, contract farming, and value-chain financing ▪ Regulations and food standards to promote healthy, sustainable diets
PUBLIC SECTOR FINANCE	▪ Public expenditure reviews with an expanded focus on SDGs, particularly those related to nutrition, poverty, environmental sustainability, and employment ▪ Reallocation of resources from existing agrifood sector support to promote healthy diets and sustainable production.
CONCESSIONAL ODA AND NON-CONCESSIONAL LOANS BY MDBS AND BILATERAL AGENCIES	▪ Budget support to low- and middle-income countries to enhance fiscal resources and for reprioritization of public expenditures in support of achieving the SDGs ▪ Loans and grants to prepare specific projects and de-risk private sector investments
BANKING SYSTEM	▪ Strategic use of blended finance mechanisms to de-risk lending for agrifood system transformation ▪ Seed money to develop cooperative banking and digital platforms ▪ Unconventional monetary policies to finance specific interventions, including through "development finance" by central banks and reassessment of the role of national development banks
CAPITAL MARKETS	▪ Creation of a project preparation/incubation/acceleration facility to develop projects and investable options for private sector environmental, social, and governance (ESG) and impact investors, and other potential actors in capital markets ▪ Use of ODA, MDB lending, and public budgets to de-risk agrifood investment ventures and to leverage blended finance ▪ Digital and other institutional arrangements to link potential investors with opportunities

CHAPTER 2

Resilience
From Policy Responses to Resilient Policy Systems

JOHN McDERMOTT, DANIELLE RESNICK, AND NICHOLA NAYLOR

John McDermott is director of the CGIAR Research Program on Agriculture for Nutrition and Health, International Food Policy Research Institute (IFPRI), Washington, DC. **Danielle Resnick** is a senior research fellow, Development Strategy and Governance Division, IFPRI, Washington, DC. **Nichola Naylor** is a research fellow, London School of Hygiene and Tropical Medicine, London.

KEY MESSAGES

- Policy responses across countries have followed a similar course – beginning with lockdowns and health measures, then shifting toward fiscal, monetary, and social protection interventions – as the pandemic progressed.
- Policymakers must balance critical trade-offs among policy actions and spending priorities for health, food systems, and economies. This requires a multisectoral perspective and clearly defined values.
- Rapid policy actions must build on existing systems; therefore, the quality of established policy and delivery systems is an important determinant of success.
- Three broad features of policy systems – adaptability, coherence and coordination, and efficacy – contribute to the success of public policies.
- Policy system efficacy requires state capacity for administrative efficiency and appropriate enforcement, as well as citizen trust in government.
- Rolling out and scaling up innovations during crises depends on an underlying business environment that enables decision-making and experimentation in partnership with the private sector.

RECOMMENDATIONS

- Build effective and resilient policy systems and programs to enable decision-makers to respond to future crises in an informed, timely, and cohesive manner and in a way that builds credibility and confidence among citizens.
- Increase understanding of the interplay of health, economic, and social policy actions, gather more data, and review experience to aid decision-making.
- Develop robust public systems for the poor and vulnerable, such as social protection, nutrition, and education programs that can be quickly scaled up and adapted.
- Support an enabling business environment to spur private innovation in food systems, particularly by small and medium enterprises.
- Develop processes for policy coordination and increase capacity of policymakers to work across multiple sectors, vertically, horizontally, and temporally, to address crises and support transformation toward more resilient food systems.

COVID-19 has presented a dramatic challenge that requires a rapid and far-reaching response by governments. Most countries have taken action with new or revamped policies to address the spread of the coronavirus and its impacts, but their responses have been complicated by limited funds, capacity, and lack of experience with such a crisis. As the pandemic and the associated economic recession evolve, policymakers must balance critical trade-offs among policy actions for health, food systems, and economies. Government actions to control virus transmission have disrupted the movements of people and goods, services, and remittances – globally, regionally, nationally, and locally – and consequently disrupted economies and food systems. For the world's poor and vulnerable, particularly those who rely on labor for their livelihoods, the impacts have been severe (see Chapters 3, 5, and 6). In this chapter, we describe the evolution of policy actions taken over the past year to address these health and economic crises; examine the interconnections and balance of policy actions across public health, food systems, and the wider economy; and discuss how the world can build more resilient systems moving forward.

EVOLUTION OF POLICY RESPONSES TO COVID-19

As COVID-19 spread around the globe in the early months of 2020, many countries initiated an emergency response intended to limit viral transmission, reduce economic impacts on vulnerable groups, and protect financial and corporate activity.[1] Figure 1 depicts the evolution of public health, food system, and economic policy actions, using a graph of disease occurrence over the first year of the pandemic linked to key policy actions taken by six representative low- and middle-income countries (LMICs) (Box 1 provides further description). For the sake of simplicity in our discussion here, we break the evolution of pandemic policy responses into three phases: emergency response, recovery, and resilience. Emergency response measures have focused on urgent interventions such as health measures to control coronavirus transmission and rapid support to severely affected households and businesses. In the recovery phase, additional short-term analysis of impacts supports more targeted policies and sectoral investments as well as adjustment of health measures and implementation of vaccination plans. The rebuilding phase should develop longer-term policies and strategies for sustainability and resilience of food, economic, and health systems. However, the actual pattern of both disease and policy evolution is more complicated, with substantial overlap in implementation of these phases, which is not surprising given the unpredictable nature and persistence of this coronavirus. Most countries are still in the response and recovery phases.

In the initial emergency response phase, rapid action was essential to address the fear and pessimism of individuals and institutions, and included lockdowns to limit deaths and avoid overwhelming health systems, as well as economic stimulus measures to address the loss of incomes. Domestic and international financial resources were mobilized quickly in many countries. While the hope was that these emergency responses would eliminate or greatly reduce coronavirus infections, this effort failed in most countries. Infection spread has followed some predictable patterns linked to population density, demography, and seasons, but these are invariably complicated by hard-to-predict waves of infection and "super-spreading" behaviors and events.[2] As a result, initial emergency health responses were adapted during subsequent disease waves.

As economic, food security, and social challenges of the pandemic became apparent, countries began to adopt short-term economic and financial measures to facilitate a medium- and long-term pandemic response that would spur recovery and promote resilience. Social protection and other programs to support vulnerable populations were expanded or put in place over the course of the year. Near the end of 2020, attention began shifting to recovery measures, such as planning for vaccinations.

The unprecedented nature of the pandemic has meant that policymakers had little evidence to draw on for choosing policies during initial stages. Impressive efforts started early and continue to collect, collate, and analyze a wide range of pandemic-related policy actions taken by different countries; these are making a wealth of information available to policymakers. One widely used policy aggregator is the Oxford Global Directory for COVID Policy Trackers and Surveys.[3] Many of these efforts aim to compare and contrast country policy experiences, but such comparisons are challenging given the range of country contexts, resources, and capacities for responding to the crisis.[4] In general, policymakers relied on broad principles and guidance, which they adapted to the evolving national and subnational conditions they faced,

BOX 1 COUNTRY CONTEXT, CORONAVIRUS INFECTIONS, AND POLICY RESPONSES

The policy actions taken by the countries included in Figure 1 reflect their unique circumstances, infection dynamics, and societal disruptions. **Bangladesh**, a densely populated, lower-middle-income country, faced an acute double challenge of coronavirus transmission and loss of livelihoods for its many poor day-laborers. Strict lockdowns were not accepted and could not be enforced, but COVID-19 cases remained relatively stable. Without large-scale social protection infrastructure, financial stimulus was the main policy response. Neighboring **Myanmar** was able to implement an initial temporary lockdown that limited cases, but it had to be reinstated in a major city when a new wave of infection occurred. **Egypt**'s main challenges were the loss of tourism and remittance incomes and its effect on households and businesses. Strict lockdowns were avoided and COVID-19 cases remained relatively controlled, allowing policymakers to focus on support to households and businesses and to consider longer-term measures to enable efficiency, effectiveness, and competitiveness of food systems (see Box 2). In **Rwanda**, social compliance with rules and regulations is high, and the country managed to control infections with targeted lockdowns and move quickly to economic recovery. **Zambia** also used targeted lockdowns while implementing social protection programs; however, the country faced a marked second infection wave late in 2020, and a government debt crisis will limit subsequent stimulus. Finally, **Honduras** implemented strict lockdown measures but other concurrent crises, including hurricanes Eta and Iota, overwhelmed an already fragile healthcare system. Emergency social measures focused on household food supplies. See also comparable figures in the Regional Developments section of this report.

FIGURE 1 Evolution of disease dynamics and policy actions in 6 low- and middle-income countries (March 2020–January 2021)

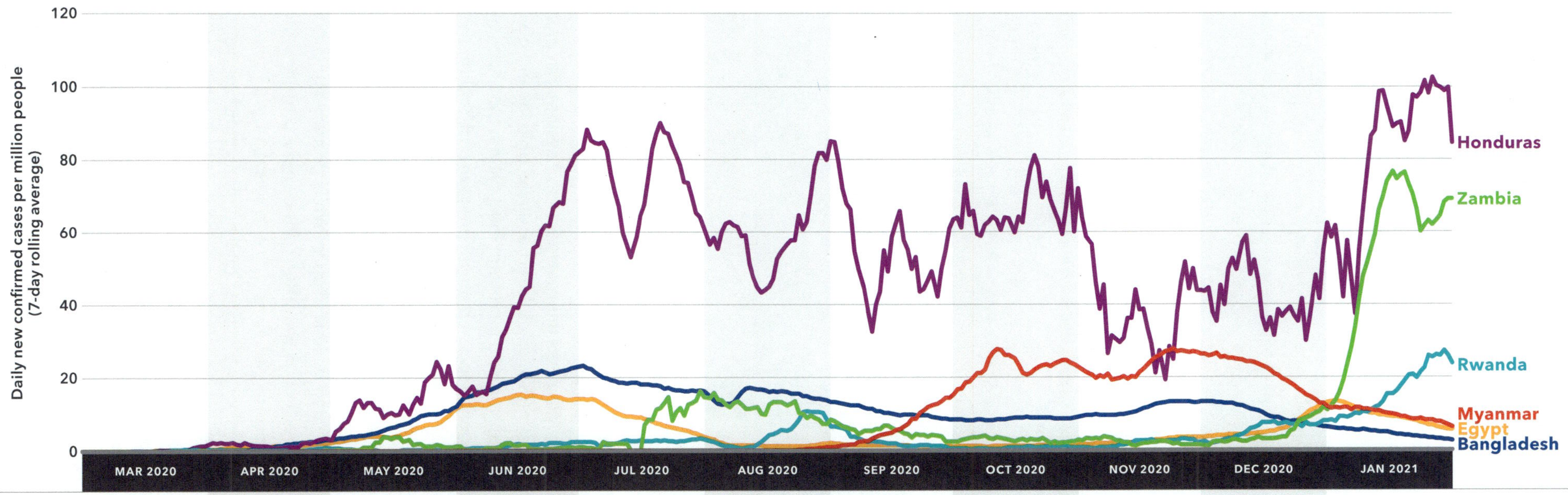

LOCKDOWNS

- **Bangladesh**, **Honduras**, and **Rwanda** issue lockdowns
- **Myanmar** issues stay-at-home order
- **Zambia** implements targeted lockdown of Nakonde
- **Rwanda** imposes targeted lockdown in Kigali
- **Myanmar** announces stay-at-home orders in Yangon

STIMULUS AND FINANCIAL ASSISTANCE

- **Egypt** issues $6.25 billion package
- **Bangladesh** issues $8 billion stimulus
- **Rwanda** receives IMF loan
- **Bangladesh**, **Egypt**, **Honduras** receive IMF loans
- **Myanmar** receives IMF loan
- **Rwanda** announces Economic Recovery Fund
- **Zambia** issues $438 million COVID-19 bond

SOCIAL PROTECTION

- Operation **Honduras** Solidaria provides 30 days of groceries to 800,000 families
- **Egypt** expands Takaful & Karama monetary subsidies
- **Rwanda** distributes food to urban poor who cannot work
- **Zambia** provides $40–$80 per month to affected households in major cities

VACCINATION PROTOCOLS

- **Bangladesh** signs deal to buy 30 million doses of AstraZeneca vaccine
- **Egypt** receives Sinopharm vaccine
- **Myanmar** secures $950 million for vaccines
- **Rwanda** submits to COVAX framework
- **Zambia** receives COVAX approval and plans to start vaccinating in February

FOOD SECTOR INTERVENTIONS

- **Honduras** offers Productive Solidarity Bonus to 190,000 producers
- **Zambia** suspends weekly markets
- **Bangladesh** closes through May
- **Bangladesh** allocates more than $1 billion to subsidize fertilizer
- **Egypt** postpones farmers' debt payments, fixes wheat price
- **Myanmar** limits hours of agriculture commodity markets
- **Myanmar** announces a $60 million plus relief plan for the agriculture sector
- **Honduras** closes 6 urban markets
- **Rwanda** allows markets to open for essential vendors, not to exceed 50% of traders
- **Rwanda** temporarily closes Kigali Market, other markets operate at 50% with traders working on rotational basis

Source: COVID-19 cases: Johns Hopkins University, CSSE COVID-19 Data (updated Feb. 19, 2021); Policies: IFPRI, COVID-19 Policy Response Portal (accessed 2021); IFPRI, COVID-19: Measuring Impacts and Prioritizing Policies for Recovery (accessed 2021).

BOX 2 POLICY PRIORITIES IN EGYPT

Dr. Ahmed Kamaly, Deputy Minister, Egyptian Ministry of Planning and Economic Development

At the outset of the pandemic, the Ministry of Planning and Economic Development had to assess the impacts on various components of GDP, including employment, poverty, and the balance of payments. To gain a comprehensive assessment, we consulted with other ministries, business associations, civil society, and academia. We tried as much as possible to rely on evidence-based policymaking because with this type of crisis, and the huge uncertainty around it, nobody knew what was going to happen. So, we turned to many sources of information, including the UN agencies, IFPRI, IMF, World Bank, and other entities.

Although we had many priorities, there were a few areas we felt needed to be especially targeted. For instance, at the beginning of the crisis, we recognized that informal workers were going to be badly affected and therefore, we provided unemployment benefits to these individuals, which we continue to do. We further anticipated a big hit in the tourism and manufacturing sectors and tried to identify policies to support those sectors. We also aimed to support ICT-based solutions because ICT is both the future and essential for mitigating the negative impacts of COVID-19 on other sectors. In addition, we provided some financial support to help businesses cover their operational expenditures and thereby prevent bankruptcies.

While it is important to track the costs versus benefits of different policy actions to determine whether to continue them, this is very difficult to do in response to COVID-19 because we adopted bundles of policies simultaneously and identifying their impacts will take time. However, the crisis has also taught us that with shocks, there are some opportunities. So, we are now looking at how we should reform our business climate and increase the productivity of our supply chains to be more competitive when this storm passes. In addition, the pandemic has highlighted that while social protection is key for helping the poor, poverty is only a symptom of the root problem, which is the lack of decent jobs in Egypt.

Note: These are personal observations and should not be construed as the official positions of the ministry.

including national public health, economic, and sociopolitical contexts. Our interviews with policymakers reflect the different perspectives they considered in the short and medium terms (Boxes 2 and 3).

Given the urgency of the pandemic, implementation of policy actions had to build on existing systems and programs; thus the quality of the established policy and delivery systems was an important determinant of success. For example, early policy responses often expanded and adapted existing social protection programs (Chapter 5). This was effective in Ethiopia, where the well-established Productive Safety Net Program was expanded.[5] While these social protection programs are relatively common in middle-income countries, there were few such programs to build on in many low-income countries. Also, the limited fiscal capacity of low- and lower-middle-income countries has meant that, after the initial emergency actions, critical decisions and trade-offs had to be made, for example, selecting which people and places to target with the limited funding available. In some countries, such as Zambia, debt burdens may preclude any additional government action.

What has this meant for food systems? In most countries, established agrifood policies and objectives supported a long-term food system transformation agenda focused on job creation and healthier and sustainable diets (as described in IFPRI's *2020 Global Food Policy Report*).[6] However, implementation lags and new issues that surfaced during the pandemic should lead to adaptations to better support recovery and build resilience. While primary agricultural production was often considered a priority sector for ensuring food security and so exempted from pandemic restrictions, food marketing and services usually were not. In the absence of targeted policies to balance restrictions on food markets and services with longer opening hours and more space, these operations suffered extensive disruption.[7] In Nigeria, for example, primary production declined by 15 percent

BOX 3 POLICY COORDINATION AND OPPORTUNITIES IN NIGERIA

Dr. Andrew Kwasari, Senior Special Assistant to the President on Agriculture and Head of the Project for Agriculture Coordination and Execution (PACE), Office of the Vice President of Nigeria

Coordination is a very important consideration for the Nigerian government as it confronts COVID-19. At the beginning of the pandemic, the President of Nigeria constituted a committee chaired by the Vice President. This committee included five sectors viewed as key for the country: agriculture, trade and investment, works and housing, minerals, and petroleum and gas. Along with the minsters for these sectors and technical support staff, an Economic Sustainability Plan (ESP) was developed that became the main framework for coordinated policy action. During the process of developing the ESP, the Vice President consulted with the Nigeria Governors' Forum. We selected one governor per geopolitical zone who then consulted with other governors in their zones, and then we had a series of virtual meetings at every stage of the ESP to ensure broad stakeholder support. We also consulted widely with commodity associations, regionally based think tanks, and captains of industry in agriculture, such as Aliko Dangote.

We realized very early on, from both our own analyses and those of international organizations and consulting firms, that agriculture could be badly affected by the coronavirus. We anticipated that there would be food shortages globally as countries prioritized their own food security over international food trade. At the same time, our main source of foreign exchange – oil and gas – was hit by low demand due to shuttered industries and a rift within OPEC (Organization of the Petroleum Exporting Countries). So, we realized that due to reduced potential for importing food, it was critical to protect our national food system and our rural populations from the virus. We decided to focus on support for local production by providing farmers with needed inputs and finance and solving logistical barriers along the food chain. For example, governors originally were stopping trucks at their state borders, but the President, Vice President, and the Federal Minister of Agriculture and Rural Development intervened to allow trucks with farm inputs and food to continue their operations. In doing so, we collaborated with the National Union of Transport Workers to ensure they confirmed trucks only with those materials could cross state borders.

Despite the economic and health costs, there have been some real opportunities due to COVID-19. For instance, we realized that if we were going to make agriculture more resilient and help entrepreneurs view agriculture as a business, we needed a database of farmers to identify where they are located and how best to support them. With financing from the federal government, technical staff from the states, and the development of the Nigerian Agricultural Assets Survey App, we trained young enumerators across the country to register farmers, collect their farm GIS information, and gather other information about the commodities they cultivate. By the end of 2020, we had 5.4 million farmers registered. We subsequently worked with a variety of commercial banks, and now 2.3 million of these farmers are financially included in the banking system. By creating this digital database of farmers, we also now have ways of depositing subsidy transfers for agricultural inputs into their bank accounts. In this way, we have leveraged different technologies to strengthen the food system to face future crises.

Note: These are personal observations and should not be construed as the official positions of the government.

and food services by 80 percent.[8] Trade and marketing of cereals and other staples were relatively unaffected, although some standard shock responses, such as restricting cereal exports (for example, rice from Viet Nam), were implemented temporarily. Supply chains for perishable products often suffered the greatest disruptions, in part because of concerns over the safety of animal-sourced foods.[9] On the positive side, these disruptions also accelerated private sector innovation, particularly for digital tracking and service delivery, that can contribute to food system transformation (Chapter 6).[10]

THE CHALLENGE OF BALANCING PUBLIC HEALTH, FOOD SYSTEM, AND ECONOMIC IMPACTS

As the pandemic led to widespread disruptions across the whole of society, the typical standalone sectoral policy responses used to address financial or food-supply crises or disease outbreaks, such as SARS and Ebola, were inadequate – more cross-governmental and cross-sectoral policy design and implementation were needed.[11] The health and economic trade-offs associated with pandemic policy decisions are not simple.[12] On the health side, policies are needed to prevent widespread illness and death, high healthcare system costs, and subsequent lost productivity. Uncontrolled virus spread can overwhelm health systems, with estimated costs ranging from US$50 per capita in low-income countries to $84 in upper-middle-income countries.[13] Thirty-day nationwide lockdowns (which aim to reduce transmission, illness and deaths, and healthcare needs) of various stringencies are thought to reduce the healthcare burden substantially, from an estimated average of $5.2 billion down to $4.7 billion per country.[14] However, lockdowns may also impose indirect health losses through reduced access to care and a heavy economic burden, especially on the most vulnerable in society, through lost income (Chapter 5).[15]

On the economic side, the pandemic together with the responses designed to slow its spread have caused a global recession. Economic growth is expected to have fallen by 0.9 percent in 2020 for low-income countries as a result of the pandemic.[16] This economic contraction is linked to multiple disruptions, both domestic and international,[17] which have sparked great concern about the pandemic's trajectory, the long-term impacts on growth and other economic indicators, and the future economic landscape.[18] Thus the implications of coronavirus control and eradication extend beyond boundaries, requiring a global solution.

To illustrate the association between lockdowns and reduced economic activity, Figure 2 plots a snapshot of the stringency of lockdown and other closure measures for 14 countries in the early months of the pandemic against the declines in national GDP and agricultural GDP (relative to the same period in 2019). In general, more stringent lockdowns are associated with greater decreases in GDP (with some exceptions, such as Mali and Malawi). However, the decline is not uniform; lockdowns have different impacts on different sectors. The association is generally weaker for agricultural GDP, likely reflecting the fact that agriculture was exempted from some restrictions. For low-income countries, primary agriculture, which suffered less disruption than value-added food sectors beyond the farm, is a much greater component of overall food sector GDP. Thus, lockdowns would be expected to have less impact on their agricultural GDP. For higher-income countries, where a greater share of food sector GDP is generated beyond primary agriculture, lockdowns may have a greater impact. They will need to be considered more carefully in the policymaking process in order to better understand the trade-offs being made in fiscal decisions and in burden-sharing as food systems transform.

The distribution of the disease burden and the economic burden across social groups within countries is also an important concern for policymakers. For example, in Bangladesh, a strict lockdown was not enforceable for poor day-laborers. Throughout low- and middle-income countries, the poor and other vulnerable populations bear the brunt of the economic impacts of response policies and are also less able to avoid contracting the virus.[19] For instance, disproportionately more people in lower socioeconomic quintiles in Pakistan and the Philippines experienced income loss and were unable to socially distance in lockdown simulation exercises.[20]

To mitigate the health and economic impacts on vulnerable populations, a World Bank policy framework advocates for a fast, hard, and short response to stop the transmission of the virus, followed by substantial fiscal and monetary stimulus in a recovery phase to mitigate the economic impacts.[21] However, as discussed above, LMICs may not have capacity for the fiscal and monetary stimulus necessary to adequately support vulnerable populations. It is estimated that additional donor aid of approximately $5 billion (about half the total required) is needed to meet this shortfall and prevent millions from being pushed into poverty and food insecurity.[22] At the same time, the anticipated long-lasting economic effects of the pandemic suggest that the public may not be satisfied with just a short-term response and could expect mitigation measures to continue for the foreseeable future.

Public health policymakers are already attempting to bring economic, social, and health perspectives into their planning for recovery and future resilience to ensure well-balanced policies. Box 4 uses the example of vaccine distribution to illustrate some approaches to prioritizing public health interventions, considering different combinations of health, social, and economic parameters. With growing evidence on the role of household characteristics, behavior, capacity, and resilience in shaping outcomes during pandemics,

FIGURE 2 Lockdown stringency and change in GDP and agricultural GDP

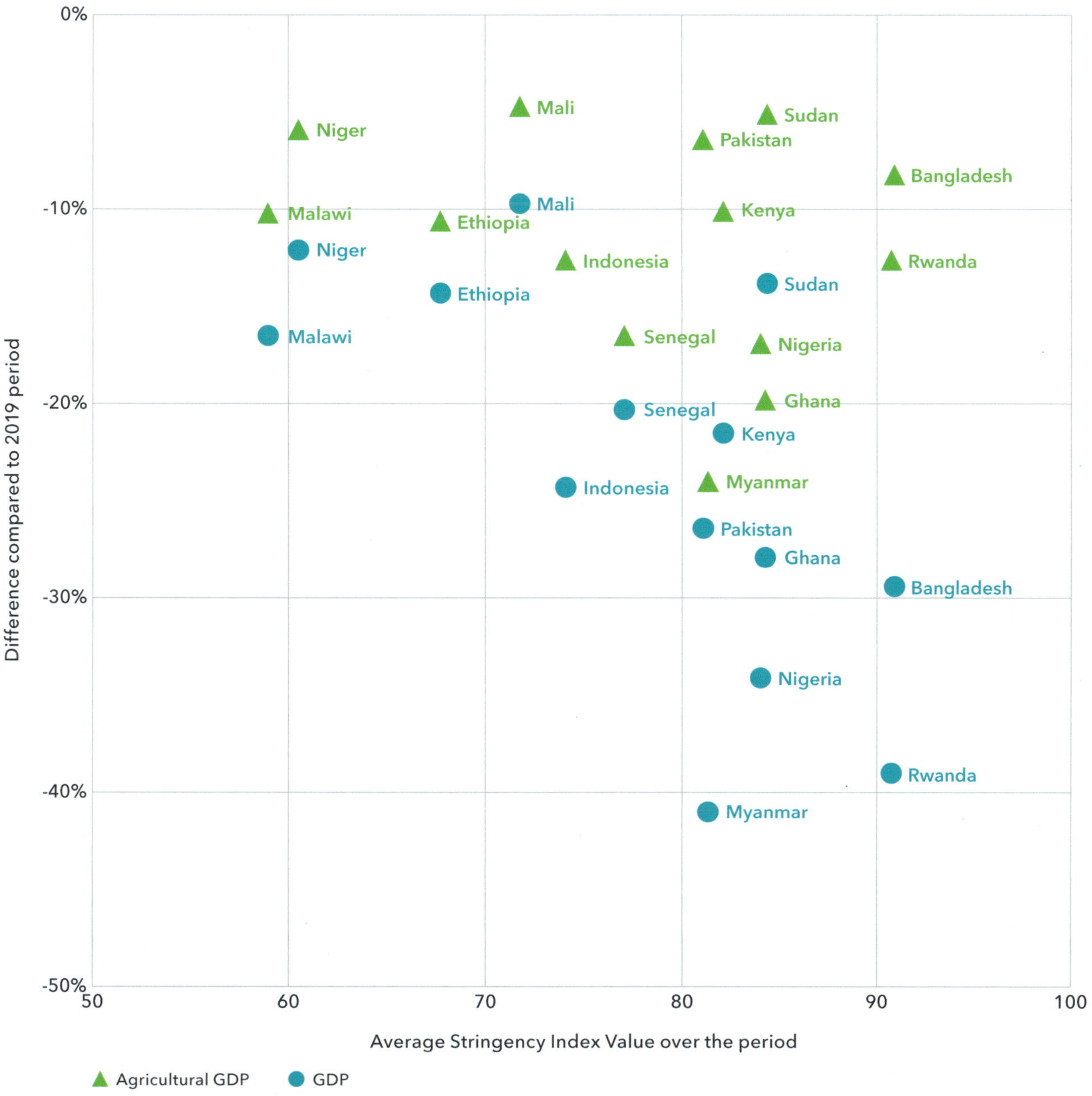

Source: Stringency data are from H. Thomas et al., Oxford's COVID-19 Government Response Tracker, Blavatnik School of Government (www.bsg.ox.ac.uk/covidtracker, accessed Feb. 8, 2021), and averaged over the same time period as the snapshot analysis of GDP and agriculture- and food-related GDP performed by IFPRI (see COVID-19: Measuring Impacts and Priotizing Policies for Recovery).

Note: Stringency is measured by the Oxford Stringency Index (national level), which is a composite measure of nine policy response indicators such as "restrictions on internal movement." Declines in GDP (gross domestic product) and agricultural GDP are relative to the average corresponding values in the same period in 2019 (for example, if lockdown occurred in March–April in 2020, the GDP for that period was compared to that of March–April 2019). The dates of the analyses vary by country due to different start and end dates of national policies, and dates of latest available economic analyses. GDP snapshot periods run from a few weeks to a few months between February and June 2020. Some countries included had a full lockdown, others had partial lockdowns or curfews, which is reflected in the stringency index.

BOX 4 INCORPORATING ECONOMIC AND SOCIAL PARAMETERS INTO HEALTH DECISIONS: TARGETING VACCINE DISTRIBUTION

In deciding whether to implement a national lockdown, what testing protocols to implement, or whom to vaccinate, a country is making value choices about what objectives to prioritize and how to distribute the burden of COVID-19 and the costs and benefits of associated policies across the population. Taking different ethical and social justice perspectives will lead to prioritizing different policies and populations. Here we explore how different perspectives can shape vaccine distribution.

ONE HEALTH PERSPECTIVE: Different decision-makers across the One Health[a] system may have different, and sometimes even competing, perspectives and objectives that emphasize human health, animal health and use, agriculture, environment, and other government goals.[b] For example, national treasuries may be interested in minimizing short-term economic GDP impact while environmental agencies may be interested in preventing longer-term pollution. Explicitly defining and understanding these differences early can allow for more effective discussion, selection, and implementation of COVID-19 vaccination strategies.

SOCIAL JUSTICE PERSPECTIVE: Deciding whom to vaccinate first presents an ethical dilemma when vaccines are scarce. Should countries seek to maximize benefits across the whole population (which usually means favoring the young, who have more years of life ahead of them than the elderly) or seek to minimize inequality and inequity (by targeting those in greatest need or with the least access)?[c] By clarifying the aggregate costs in terms of years of life lost or the aggregate GDP impact of specific vaccination policies, along with the distribution of these impacts across specific subpopulations, both overall benefits and equity impacts can be considered in a structured way.[d] For example, for infectious diseases such as COVID-19, poorer populations may be more severely affected since they have limited capacity to socially distance and only obtain income through in-person labor. Thus vaccinating such frontline poor workers mitigates multiple risks and has multiple benefits.

The World Health Organization SAGE Working Group on COVID-19 vaccines proposes that policymakers allow for "the integration of explicit values with evolving scientific and economic evidence."[e] Recommended values to consider include human wellbeing, equal respect, global equity, national equity, reciprocity, and legitimacy. Building on this, scenarios for various levels of vaccine availability are proposed:[f]

- 11–20% of national population: Sociodemographic groups at significantly higher risk of severe disease or death should receive the vaccine.
- 21–50% of national population: Extend vaccination to essential workers outside the health and education sectors, including agriculture and food workers and groups at elevated risk of acquiring and transmitting infection because they are unable to effectively physically distance.

National prioritization decisions may select different priority populations based on the levels of risk of infection or harm (in all senses described in this section) and on the preference given to different values within that country.

Source: [a]"One Health is a collaborative, multisectoral, and transdisciplinary approach – working at the local, regional, national, and global levels – with the goal of achieving optimal health outcomes recognizing the interconnection between people, animals, plants, and their shared environment" (CDC, accessed Feb. 2021, https://www.cdc.gov/onehealth/basics/index.html); [b]N.R. Naylor et al., "Quantitatively Evaluating the Cross-Sectoral and One Health Impact of Interventions: A Scoping Review and Case Study of Antimicrobial Resistance," *One Health* 11 (2020): 100194; [c]A. Giubilini, J. Savulescu, and D. Wilkinson, "COVID-19 Vaccine: Vaccinate the Young to Protect the Old?," *Journal of Law and the Biosciences* 7, no. 1 (2020): lsaa050; [d]R.E. Glover et al., "A Framework for Identifying and Mitigating the Equity Harms of COVID-19 Policy Interventions," *Journal of Clinical Epidemiology* 128 (2020); [e]World Health Organization, *WHO SAGE Values Framework for the Allocation and Prioritization of COVID-19 Vaccination* (Geneva: 2020); [f]S. Omer et al., *WHO SAGE Roadmap for Prioritizing Uses of COVID-19 Vaccines in the Context of Limited Supply* (Geneva: World Health Organization, 2020).

along with the expanding pool of disease-transmission models and associated micro- and macroeconomic costing models, 2021 should see more deliberative, evidence-based policymaking that recognizes both the health and economic costs and benefits of policy options for different sectors and social groups.[23]

TOWARD EFFECTIVE AND RESILIENT POLICY SYSTEMS

As countries move into the recovery and rebuilding stages, we must think about how to make policy systems more effective and resilient. Just like households, policy systems may not be able to withstand shocks and can become constrained by inertia that allows crises to fester or become fragmented by inconsistent and volatile policy initiatives. "Resilient policy systems" therefore refers to systems that enable decision-makers to respond to future crises, single and multiple, in an informed, timely, and cohesive manner and in a way that builds credibility and confidence among citizens. While policy interventions for the current pandemic have varied greatly across countries, certain features of policy systems have been shown to enhance the ability to make rapid and efficacious decisions vis-à-vis health, food systems, and economic livelihoods under conditions of uncertainty. Three broad features of policy systems discussed here – adaptability, coherence and coordination, and efficacy of implementation and enforcement – contribute to both pandemic recovery and future food system transformation.

ADAPTABILITY refers to the ability to modify extant policies to meet new needs, arising either from crises or longer-term changes. In 2020, some countries quickly found innovative ways both to keep markets working while reducing disease transmission and to refocus public healthcare delivery to address pandemic conditions. In Sri Lanka, the century-old weekly tea auction was quickly shifted to an online platform. In East Africa, the Regional Electronic Cargo and Driver Tracking initiative was introduced in response to the pandemic and has facilitated the monitoring of truck drivers for COVID-19, reducing congestion and long wait-times at borders.[24] In Ghana, drone technology was deployed in rural areas to speed the transport of medical supplies and medical samples for coronavirus testing.[25] In a number of countries, urban informal workers were incorporated into social protection programs as lockdowns closed markets or reduced trading hours.[26]

Many of these quick adaptations could be successfully implemented under emergency conditions because they built on established programs and relationships between societal groups and the state. The ability to roll out and scale up some innovations during crises depends on an underlying business environment that enables decision-making and experimentation in partnership with the private sector. For example, market innovations and public interventions that rely on information and communications technologies (ICTs), such as digital cash transfers, are more likely to succeed when enabled by established technology governance that encourages innovation and cooperation (without generating new societal costs with respect to privacy, misuse of data-sharing, or inequities in access).[27] Existing organizations and negotiating processes can also be leveraged quickly for implementation. For instance, one survey found that disbursement of pandemic cash relief payments to informal workers was more effective in cities where active informal workers' associations could help members navigate digital registration platforms.[28]

COHERENCE AND COORDINATION are critical features of policy systems that mitigate against volatility, confusion, and inefficient outcomes. Such coordination has at least three different dimensions: horizontal (across sectors), vertical (across levels of government), and temporal (sequencing of policies). Many countries initially focused on ***horizontal coordination*** and established interministerial response teams; from a sample of 33 countries followed by IFPRI's COVID-19 Policy Response Portal, 23 created a COVID-19 taskforce or committee at the start of the pandemic that typically included ministries of health, trade, transport, foreign affairs, police, tourism, finance, and defense.[29]

Vertical coordination was essential given that local governments have played frontline roles during the pandemic.[30] In Uganda, local governments are responsible for surveillance, behavioral education to limit contagion, enforcing control measures such as curfews, identifying beneficiaries for food distribution, and delivering food aid.[31] The Kampala Capital City Authority even launched an online app for the home-delivery of food from informal vendors whose livelihoods were threatened by the lockdowns.[32] However, vertical coordination is especially difficult in countries with a high degree of political polarization or with substantial subnational political autonomy, such as federal countries with decentralized health systems. In Nigeria, for example, the governors of two states rejected the testing guidelines of the Presidential Task Force.[33] In contrast, in Brazil, governors adopted more stringent coronavirus lockdowns and testing and lobbied for quicker vaccination options than the country's president thought necessary.[34]

Temporal coordination has proved exceedingly difficult. In India, the president announced lockdowns with only four hours' notice; but transport systems were unprepared to adapt quickly to meet the needs of millions of migrants returning to rural areas, contributing to a surge in coronavirus infections.[35] In many African cities, markets were shuttered to assist with social distancing, but this health measure also deprived local governments of needed revenue for water and sanitation management, which are necessary for effective hygiene.

POLICY EFFICACY depends on implementation and credible enforcement and requires state capacity – in terms of both enforcement abilities and administrative efficiency – as well as citizen trust in government. Administrative efficiency depends on the state's technical competency and skills as well as adequate equipment for implementation. This form of capacity is essential to the rapid rollout of any emergency program. In settings of low administrative efficiency, the initial flurry of cash-transfer and food-relief policies was often overshadowed by cases of government corruption, hoarding, or poor accessibility for intended beneficiaries.[36] Capacities are generally higher where there are more human and financial resources, mechanisms for oversight and accountability, and autonomous public institutions. Such independent public institutions play an especially important role in counterbalancing the substantial political discretion that characterizes rapid responses to crises and can lead to corruption and discrimination that subvert intended policy objectives.

Enforcement needs to be employed with caution. On the one hand, use of the police, military, and surveillance was essential to enforce lockdowns, mask mandates, social distancing, and contact tracing. In Africa, it may be that only Rwanda and South Africa could exert sufficient enforcement capacity to maintain effective lockdowns, and these two countries indeed had among the most stringent lockdowns in the region.[37] On the other hand, where deployment of enforcement capacity is excessive, it can lead to human rights abuses and stifle the free flow of information, raising questions about government motivations.[38]

Administrative and enforcement capacities may interact.[39] For instance, in Bangladesh, citizens first complied with the government's lockdown, which involved daily deployments of the military and police to vacate streets after 2 p.m. and close all shops that were subject to the lockdown. But, as the state failed to deliver promised economic assistance and citizens became economically desperate, they began to defy restrictions. In response, the police slowly abandoned the enforcement of lockdowns in May, first tacitly and then officially, despite rising COVID-19 cases.[40]

While balanced but effective state capacity is key for policy efficacy in the short-term, trust in evidence-based information from government, media, or social networks can also be a critical determinant of citizens' compliance with health measures over time.[41] These processes can be mutually reinforcing; for instance, misinformation exacerbating distrust in scientific evidence related to the coronavirus, or disappointment related to a government's handling of other pandemic-related policies, can create skepticism about the efficacy and safety of coronavirus vaccinations. Many governments now recognize that building trust is fundamental for citizen acceptance of vaccinations and are identifying strategies to do so.[42] These include engaging respected community leaders, empowering frontline health staff, creating forums for citizens to share concerns, identifying differential concerns across gender and ethnicities, and leveraging multiple information platforms.[43]

POLICIES FOR RESILIENT FOOD SYSTEMS

In sum, the coronavirus pandemic has been an unusually great challenge for policymakers because it has simultaneously disrupted healthcare systems, food systems, and economic systems. As the impacts of

COVID-19 have shifted over time, policy responses have likewise evolved based on country contexts and capacities. Balancing health and economic policy actions has been difficult, but there is increasing understanding of their interplay and increasing data and experience available to aid decision-making. Finally, although there is no standard policy framework that all countries can apply, building an effective and resilient policy system characterized by adaptability, coherence and coordination, and efficacy of implementation and enforcement can greatly facilitate policy responses and contribute to the development of more robust and resilient food systems.

Prior to the pandemic, food system transformation was an important development strategy of most LMICs. The pandemic has exposed vulnerabilities in food system transformation strategies – particularly for the key policy objectives of inclusion and jobs for poor and disadvantaged people and more diverse food supplies, especially of nutrient-dense perishables. Two types of longer-term policy actions and investments have proved important during the pandemic. One is support for an enabling business environment, including flexible food trade and markets. These have spurred needed private innovation, particularly by small and medium enterprises. The second is robust public systems for the poor and vulnerable, such as social protection, nutrition, and education programs that can be quickly scaled up and adapted. While there is no common formula to address these challenges across all countries, increased capacity to consider multiple sectors and build effective and resilient policy systems are common elements in supporting more effective food system transformations.

CHAPTER 3

Nutrition
Transforming Food Systems to Achieve Healthy Diets for All

MARIE RUEL AND INGE D. BROUWER

Marie Ruel is director of the Poverty, Health, and Nutrition Division, International Food Policy Research Institute, Washington, DC. **Inge D. Brouwer** is associate professor in the Division of Human Nutrition and Health, Wageningen University & Research, the Netherlands, and leader of the Food Systems for Healthier Diets flagship under the CGIAR Research Program Agriculture for Nutrition and Health (A4NH).

KEY MESSAGES

- Pre-pandemic, 3 billion people could not afford a healthy diet; that number could rise by 267.6 million between 2020 and 2022 due to the pandemic.
- Evidence from phone surveys in low- and middle-income countries shows widespread job and income losses and rapid rises in food insecurity due to government measures to contain the pandemic; poorer households, women, and other vulnerable groups are most affected.
- Across the globe, the quality of diets deteriorated due to disruptions in supply of fresh, healthy foods, drops in demand for these foods due to unaffordability and perishability, and increased consumption of cheaper sources of calories, including starchy staples and ultra-processed foods.
- Deteriorations in diet quality could have devastating consequences for the health and nutrition of vulnerable women and children, increasing all forms of malnutrition in the short term and causing lifelong, irreversible development, health, and nutrition damage, reversing decades of progress made so far.
- Food system transformation must support healthy diets and by doing so, serve as double duty actions that simultaneously tackle all forms of malnutrition.

RECOMMENDATIONS

- Strengthen and expand coverage of targeted social protection programs, including cash and food transfers, to support healthy diets through, for example, behavior change communication focused on healthy diets and lifestyle, direct incentives such as vouchers for healthy foods, and improving the quality of school meals.
- Provide support to low- and middle-income countries for developing national food-based dietary guidelines defining the minimum dietary standards to prevent all forms of malnutrition and for formulating dietary targets for public and private investment strategies.

- Harness the influence of food environments to redirect food systems toward healthier food provision that is profitable yet supportive of optimal health and nutrition, for example, by regulating advertising and marketing of unhealthy food products or encouraging better food choices via a combination of taxes and subsidies.
- Pair large-scale demand creation and behavior change communication strategies with innovations in food environments and food supply systems and policy incentives in order to maximize impacts for healthy diets.

The COVID-19 pandemic, together with the economic crisis it precipitated, has revealed the fragility of local, national, and global food systems and shed new light on their vulnerabilities and shortcomings in supporting the achievement of healthy diets for all. In recent decades, food systems have been evolving in ways that undermine healthy diets even in the best of times, as cheap, ultra-processed foods have become widely available and fresh, nutritious foods less affordable and less accessible to many – especially the poor and especially in low- and middle-income countries (LMICs).

Before the pandemic, some 3 billion people already could not afford a healthy diet.[1] Another 267.6 million are now expected to join their ranks between 2020 and 2022 due to the pandemic and resultant economic downturn.[2] The unaffordability of healthy diets is likely to hit poor people in LMICs the most because these populations tend to be more sensitive to income and

This chapter draws from I.D. Brouwer, M.J. van Liere, A. de Brauw et al., "Reverse Thinking: Taking a Healthy Perspective Towards Food Systems Transformation," IFAD Nutrition Cornerstone paper in preparation of the RDR2021, Wageningen University & Research, 2020.

food price shocks and have been disproportionately affected by pandemic-related job and income losses.[3]

Widespread food insecurity and a shift toward consumption of low-quality diets could, in turn, have devastating consequences for health and nutrition in LMICs, especially among women of reproductive age and young children.[4] Recent estimates suggest that, between 2020 and 2022, the pandemic could result in 9.3 million newly wasted and 2.6 million newly stunted children; 168,000 additional child deaths; and 2.1 million additional cases of maternal anemia.[5] It could also exacerbate already soaring rates of overweight and obesity observed in LMICs in recent decades, especially in countries in Asia and Africa that are experiencing a rapid modernization in food supply and a "nutrition transition" – that is, a shift toward diets high in sugar, salt, saturated fats, refined grains, and ultra-processed foods.[6]

In preparing for the post-pandemic era, we need to do more than rebuild our food systems: We must transform them. And transformation must go beyond achieving resilience and sustainability to also delivering healthy and affordable diets and wellbeing for all. Taking a healthy diet perspective, this chapter identifies key policy actions that need to be prioritized to ensure that food system transformation supports healthy diets and achievement of optimal health and nutrition for the growing world population.

WHAT IS A HEALTHY DIET AND WHY IS IT IMPORTANT?

A healthy diet is one that provides adequate amounts of energy and all essential nutrients and promotes optimal nutrition and health (Box 1). A healthy diet supports the immune system and is therefore particularly important during the pandemic to help reduce the risk and severity of illness from the virus.[7]

Poor diets, on the other hand, are the main driver of all forms of malnutrition, including childhood wasting and stunting, micronutrient deficiencies, adult underweight, overweight and obesity, and diet-related noncommunicable diseases. Poor diets are currently estimated to cause 22 percent of all deaths and 15 percent of disability-adjusted life years (years lost to illness, disability, or early death) among adults worldwide.[8] The burden is highest in LMICs,[9] where many people were living on suboptimal diets even before the coronavirus pandemic. Although the world's poorest persistently suffer from (periods of) hunger and

BOX 1 WHAT IS A HEALTHY DIET?

A healthy diet ensures adequacy of energy and all essential nutrients, promotes all dimensions of individual health, and prevents malnutrition in all its forms and diet-related noncommunicable diseases, such as type 2 diabetes, cardiovascular diseases, and some forms of cancer. A healthy diet includes enough fruits, vegetables, nuts, seeds, whole grains, and legumes; sufficient but not excessive calories and amounts of starchy staples and animal-sourced foods (milk, eggs, poultry, and fish); and limited or no foods, food groups, or nutrients that could lead to health risks when eaten in excess, such as free sugars (including sugar-sweetened beverages), saturated fat, salt, red and processed meats, and ultra-processed foods. A healthy diet should have only minimal levels, or none if possible, of pathogens, toxins, and other agents that cause foodborne diseases.[a]

The exact makeup of a healthy diet varies depending on individual requirements and physical activity, cultural context, local food availability and access, and dietary customs, but there are general principles for making healthy diets possible. These include ensuring that a diversity of nutritious and safe foods are available and accessible year-round; that healthy diets are affordable to all; that foods are produced with a low environmental footprint; and that consumers are informed, empowered, supported, and willing to make healthy dietary choices.[b]

Source: [a] WHO (World Health Organization), "Healthy Diet," fact sheet, April 29, 2020; [b] GLOPAN (Global Panel on Agriculture and Food Systems for Nutrition), *Future Food Systems: For People, Our Planet, and Prosperity* (London: 2020).

lack consistent access to diverse, safe, and nutrient-rich foods, other poor population groups are rapidly shifting to consuming diets with excess calories, saturated fats, salt, and sugar and foods that do not support optimal health and nutrition such as ultra-processed foods.[10] It is imperative that food systems start contributing not only to providing enough calories to feed the world, but also to supporting achievement of high-quality diets that promote optimal health and nutrition, all while having a small environmental footprint and supporting livelihoods.[11]

IMPACTS OF THE COVID-19 PANDEMIC

During the early stages of the pandemic, governments around the world implemented a series of measures to control virus transmission. Countries imposed combinations of complete or partial lockdowns, travel and migration restrictions, closure of restaurants and other food retailers, and suspension or reduction of formal and informal food sector activities for weeks or months at a time. With travel and transport restrictions, food, farm, and other workers along food value chains lost their jobs, crops went to waste, and the supply of perishable nutritious foods suffered as a result.[12] At the same time, the massive losses of employment and income decreased demand for relatively expensive nutritious animal-sourced foods and fresh fruits and vegetables, which affected poor households and women disproportionately.[13] Other factors contributing to the drop in demand for these nutritious and fresh foods, especially among the poor, included temporary price increases; lack of household refrigeration and proper storage facilities, preventing bulk purchase of perishables during lockdowns; and the misperception that fresh foods (especially animal-sourced foods) were a risk factor for virus transmission.[14]

The combination of demand and supply challenges, compounded by misinformation, led to drops in consumption of nutritious foods, shifts toward cheaper calorie sources such as starchy staples or packaged, ultra-processed foods with long shelf lives, and an overall deterioration in diet quality.[15] Global projections showed consumption of nutritious animal-sourced foods and fruits and vegetables falling between 6 and 9 percent and vegetable oil and sugar consumption rising.[16] Although some of these dietary shifts may prove to be temporary, they may have consequences for vulnerable populations (the poor, migrants, refugees, women, young children, adolescents, the elderly, and socially excluded) for years to come.[17]

Early assessments find that restriction-related job and income losses in countries in Africa south of the Sahara (Ethiopia, Kenya, Malawi, Nigeria, and Uganda) and in South and Southeast Asia (Bangladesh, Myanmar, and Viet Nam) have affected 65 to 95 percent of populations, depending on context (country, rural/urban), income sources (farm/nonfarm, wages, remittances), initial poverty levels, and gender (Box 2).[18] Food insecurity, both moderate and severe, also increased sharply at the onset of the pandemic in all countries for which data are available.

Rising food insecurity led to a series of food- and diet-related coping strategies among households, including slashing spending on food; reducing consumption of relatively expensive fresh and nutritious animal-sourced foods and fruits and vegetables, worsening dietary quality in the process;[19] and consuming fewer meals each day, sometimes even going without eating for an entire day.[20]

The dramatic increases in food insecurity and risky coping strategies are expected to exacerbate undernutrition, especially in mothers and young children.[21] In some contexts, especially in lower-middle-income and urban areas where ultra-processed foods are widely available and relatively cheap, the pandemic may lead to shifts toward over-consumption of unhealthy foods at the expense of more nutritious foods, triggering a rise in rates of overweight and obesity and exacerbating micronutrient deficiencies.[22]

Although long-term impacts of the pandemic are expected, especially as the economic recession lingers, preliminary evidence from Ethiopia and Bangladesh suggests that some poor populations may be able to bounce back soon after government restrictions are lifted, especially if they are able to resume work.[23] Some of the mitigating measures, especially the extensive social protection programs put in place in 215 countries,[24] may also help the recovery (Chapter 5). Ultimately, however, demand may exceed the capacity of such programs, their duration could be too short, or distribution could fail to reach the poorest and most vulnerable.[25]

BOX 2 ASSESSING THE IMPACTS: RESULTS FROM PHONE SURVEYS

Although results from extensive household surveys in LMICs are not yet available to document the pandemic's impacts on income, food security, diets, or nutrition outcomes, phone surveys carried out in several countries provide some insights. Most studies were conducted during or soon after government restrictions were put in place, but a few used repeated or high-frequency surveys through the second half of the year.[a]

These surveys found that the pandemic was affecting populations differently and to different degrees depending on who they were, where they lived, how they earned their incomes, and how well-off they were before. For example, in Bangladesh, income losses were more widespread among those living in urban slums than in rural areas and also among poorer groups,[b] and overall up to 84 percent of previously nonpoor households reported lower earnings than in 2018/19.[c] In Ethiopia, income losses were greatest among the poorest women.[d] A four-country study using high-frequency surveys in Ethiopia, Nigeria, Malawi, and Uganda found that 77 percent of the population lost income due to the pandemic, but no differences were found between urban and rural areas.[e] In Myanmar, monthly phone surveys showed that income poverty had already risen above pre-pandemic levels by June and continued to grow worse through October, disproportionately affecting rural populations.[f]

Similarly, pandemic-related increases in food insecurity varied in different contexts. In Africa, the four-country study showed that up to 61 percent of the adult population (or 100 million adults) had experienced moderate or severe food insecurity, with women and poorer households being hardest hit.[g] In Nigeria, larger increases in food insecurity were found among households that relied on nonfarm businesses, were poorer, and/or lived in remote and conflict-affected zones.[h] Farmers in Kenya and Uganda, meanwhile, were less likely to experience large changes in food insecurity than populations that depended on market sources for food.[i]

Different populations likewise adopted different food or diet-related strategies to cope with rising food insecurity. Households in Bangladesh, Myanmar, and Ethiopia reduced total food expenditure by 25 percent on average, with larger declines in Bangladesh among slum dwellers than among rural populations.[j] One of the most consistently reported dietary shifts was a drop in consumption of nutrient-rich animal-sourced foods and fruits and vegetables, leading to deteriorations in diet quality and increased risks of micronutrient deficiencies.[k] In Myanmar, for example, low dietary diversity became more common among women, rising from 30 percent in June/July to 53 percent in September/October 2020;[l] and in Nigeria, total food consumption (including fruits) declined due to dips in purchasing power.[m] Increased consumption of unhealthy processed snacks and fast foods was also found in some contexts, including in urban areas of Nigeria and Viet Nam in the first few months of the pandemic,[n] and there is anecdotal evidence of similar dietary shifts in Asia and Latin America.[o]

Although quantitative evidence from in-person surveys of representative population samples is still unavailable, findings from phone surveys provide useful snapshots and highlight the urgent need for direct and immediate support to poor households around the globe – and for women and young children in particular – in order to prevent long-term, irreversible health and nutrition consequences for current and future generations.

Source: See endnotes on page 110.

TRANSFORMING FOOD SYSTEMS TO SUPPORT HEALTHY DIETS FOR ALL

The risks of losing decades of progress on nutrition to the pandemic and the related economic crisis are substantial. The situation therefore calls for immediate action to prevent further deterioration in diets and to protect nutrition and health now and in the long term. Here, we focus on priority actions to shift food environments (the interface between consumers and food systems; see Box 3) and consumer demand toward healthier diets because both have received too little attention in agriculture and food system strategies until recently.[26] The policies we propose would also serve as "double duty" actions because they focus on achieving healthy diets, which is the most critical strategy to simultaneously tackle

BOX 3 WHAT IS THE "FOOD ENVIRONMENT"?

The food environment is defined as the "collective physical, economic, policy, and sociocultural surroundings, opportunities and conditions that influence people's food and beverage choices and nutritional status." The food environment is the "interface" between the food system and the consumer. It is shaped by the food system, and in turn the characteristics and policies of the food environment directly affect consumer food choices and diets. Key food environment drivers of food choices include prices and other economic factors; food availability, quality, taste, and convenience; and food promotion, marketing, labeling, and safety.

Note: Definition drawn from B. Swiburn, G. Sacks, S. Vandevijvere, S. Kumanyika, T. Lobstein, B. Neal, S. Barquera, S. Friel, C. Hawkes, B. Kelly, et al. "INFORMAS (International Network for Food and Obesity/Non-communicable Diseases Research, Monitoring and Action Support): Overview and Key Principles," *Obesity Reviews* 14, no. S1 (October 2013): 1–12.

all forms of malnutrition, including undernutrition, micronutrient deficiencies, and overweight and obesity.[27]

These actions are needed to complement renewed efforts to make food systems more resilient, sustainable, and inclusive through actions on the supply side, including strategies to make nutritious foods and healthy diets more accessible, affordable, and safer. We focus on four broad categories of policy priorities for LMICs: (1) social protection policies that support healthy diets; (2) policies to shape national dietary guidelines; (3) policies to make food environments more supportive of healthier food choices; and (4) policies to shift food demand toward healthier diets.

SOCIAL PROTECTION POLICIES THAT SUPPORT HEALTHY DIETS: Targeted social protection programs, including cash and food transfers, have been used extensively to respond to the immediate needs of populations affected by COVID-19.[28] These types of programs could be strengthened to support healthy diets more directly by including information, behavior change communication, and promotion of nutritious foods; providing direct incentives such as vouchers for nutritious foods; or offering healthy school meals (with or without a community/school garden component) or healthy meals in factories or office canteens and other institutional catering systems. The coverage of these programs should be expanded to absorb the large numbers of "new poor" resulting from the pandemic and to effectively target women and children and other vulnerable groups.[29]

POLICIES TO SHAPE NATIONAL DIETARY GUIDELINES: National food-based dietary guidelines are needed to characterize healthy diets in a culturally appropriate way for diverse contexts and population groups. However, many countries, especially in Africa south of the Sahara, do not have such guidelines; and where guidelines do exist, they are often underutilized, too vague, incompatible with health targets, or lack sustainability considerations.[30] Governments in LMICs need support in developing (or revamping) their national food-based dietary guidelines to define the minimum dietary standards to prevent all forms of malnutrition and in formulating dietary targets for public and private investment strategies.[31]

POLICIES TO MAKE FOOD ENVIRONMENTS MORE SUPPORTIVE OF HEALTHIER FOOD CHOICES: Food environments, the informal and formal places where consumers acquire their food and meals, have an enormous influence on diets, as shown by how effective they have been at triggering excessive consumption of packaged ultra-processed foods, beverages, and snacks in LMICs.[32] The influence of food environments should be used to reverse these trends and redirect food systems toward healthier food provision that is profitable yet supportive of optimal health and nutrition.[33] Examples of high-potential interventions to achieve these goals are mandatory or voluntary food labeling designed to inform consumers and to ensure consumer comprehension and use in LMICs;[34] regulation of advertising and marketing of unhealthy food products, especially for children;[35] choice restrictions through changes in options or in the physical environment (e.g., strategic positioning and presentation of healthy versus risky foods);[36] and a combination of taxes[37] or subsidies[38] to discourage or encourage specific food choices. Experience with these approaches is growing in LMICs, but research is needed to document successes, failures, impacts, and potential for scaling up.

POLICIES TO SHIFT FOOD DEMAND TOWARD HEALTHIER DIETS: There is ample literature on the use of consumer behavior change interventions, including public awareness campaigns, online interventions (such as motivational emails and use of social media influencers),[39] m-nutrition services (services delivered via mobile phone technology and apps),[40] and social marketing (activities combining ideas from commercial marketing and social sciences to achieve positive changes in behavior),[41] and on experimental use of social norms to influence healthy eating.[42] Food choices and dietary patterns can also be improved through dissemination of recipes, chef's recommendations, and cooking classes and training. However, evidence on the impact of these strategies in influencing healthy diet choices in LMICs is extremely scarce.[43] Large-scale demand creation and behavior change communication strategies targeting children, adolescents, and adults need to be paired with innovations in food environments and food supply systems and combined with policy incentives in order to maximize impacts for healthy diets.

FROM CHALLENGE TO OPPORTUNITY

The coronavirus pandemic is a global threat to healthy and nutritious diets and could increase all forms of malnutrition now and for years to come. Damage done in the short term, especially to young children during their first 1,000 days and to adolescents, may cause lifelong, irreversible development, health, and nutrition damage, compromise their economic productivity, and jeopardize their futures.[44] There remains, however, much uncertainty about the extent of devastation that the virus will leave in its wake. The ultimate damage to humanity will depend on the pandemic's trajectory and on the ability of governments to mitigate impacts and to make food, health, social protection, and education systems stronger and more resilient. Efforts to transform our food systems in the aftermath of the pandemic offer an opportunity to embed actions to refocus food environments and reorient consumer behavior toward healthier dietary choices and improve health and nutrition globally.

"It is imperative that food systems start contributing not only to providing enough calories to feed the world, but also to supporting achievement of high-quality diets that promote optimal health and nutrition, all while having a small environmental footprint and supporting livelihoods."

CHAPTER 4

Natural Resources and Environment
Governance for Nature-Positive Food Systems

RUTH MEINZEN-DICK, CLAUDIA RINGLER, WEI ZHANG, AND CHANNING ARNDT

Ruth Meinzen-Dick is a senior research fellow, **Claudia Ringler** is deputy division director, **Wei Zhang** is a senior research fellow, and **Channing Arndt** is division director, all with the Environment and Production Technology Division, International Food Policy Research Institute (IFPRI), Washington, DC.

KEY MESSAGES

- COVID-19 has brought home the necessity of better integration of natural resources and ecosystems with human food systems to increase the resilience, health, and sustainability of food systems.
- Environmental degradation and climate change, in which food systems play a prominent role, are likely to increase the frequency and severity of natural disasters and may increase future pandemics, both of which cause shocks to food and health systems.
- Common agricultural practices often degrade ecosystem services such as soil fertility and natural pest control, and can contribute to greater reliance on external inputs with potential for further damage.
- Poor people are heavily dependent on natural resources for their livelihoods and are often most severely affected by environmental shocks and resource depletion.
- The vicious cycle of unsustainable resource use and environmental degradation must be replaced with a virtuous cycle of healthier food and ecosystems using approaches that improve outcomes for humans and nature.
- National laws and institutions, plus local formal and informal institutions and norms about respecting (or exploiting) nature, shape how people interact with the natural resource base and thus the outcomes for food and natural systems.

RECOMMENDATIONS

- Protect environmental health to reduce the likelihood of emergent zoonotic diseases and natural disasters.
- Foster an integrated approach by thinking in terms of "eco-agri-food systems" that encompass natural resources and ecosystem services as part of food system policies and institutions.
- Strive for "nature-positive" food systems that maintain or even restore ecosystems and the natural resources and ecosystem services on which we all depend for our food.

- Assess how laws, institutions, and mindsets affect ecosystem functions to help identify and implement effective incentives – from regulation to market-based and normative incentives – for sustainable, nature-positive food systems.
- Tap local knowledge and goals through participatory processes and develop multistakeholder platforms to support effective, inclusive governance.
- Promote landscape- and watershed-level governance to maintain complex multifunctional landscapes that increase ecosystem services.

All food systems depend on the world's stock of natural assets, including soil, air, water, and all living things, and on the range of ecosystem services they generate. Maintenance of soil fertility, nutrient cycling, pollination, biological pest control, climate regulation, and water purification are among the many ecosystem services that are vital to agricultural production.[1] Yet human activities are causing rapid and sometimes irreversible damage to these resources and services. Climate change, land degradation, biodiversity loss, groundwater mining, solid waste, and water and air pollution, in turn, disrupt human social and economic activities, especially in our food systems. These problems are compounded by growing inequities in access to natural resources and basic necessities such as water and food; by poor governance of natural resources; by a short-term focus on GDP-driven economic growth; and by underinvestment in innovations that can reduce our environmental footprint.

The coronavirus pandemic has brought home the necessity of managing our natural resources and ecosystems to increase the resilience, health, safety, and sustainability of our food systems. The novel coronavirus likely emerged from intensified human–wildlife interactions, linked to habitat loss and environmental degradation, and future pandemics are almost certain to be caused by zoonotic diseases emerging from disrupted habitats. Other global and regional shocks to food systems and human wellbeing will also be sparked by environmental degradation and climate-change-induced disruptions that lead to conflict and worsening food security, with the largest impact on poor and vulnerable populations. Meeting the Sustainable Development Goals of Ending Poverty (SDG1), Zero Hunger (SDG2), and Good Health and Well-Being (SDG3) requires action on SDG15 to Protect, Restore, and Promote Sustainable Use of Terrestrial Ecosystems. Yet the current pandemic and other crises often mean that investments in the environment and sustainability are postponed or ignored as funding is directed to immediate humanitarian needs.[2]

Expanding our concept of "food systems" to "eco-agri-food systems"[3] – to encompass natural resources and ecosystem services – can help us meet the SDGs, reduce the impact of food production, and transform our food systems in the long term. With greater investment in the integration of natural systems with human food systems, we can even move toward "nature-positive" systems that restore ecosystems and increase the provision of ecosystem services.

FOOD SYSTEMS AND ENVIRONMENTAL DEGRADATION

Underinvestment in agriculture, together with an underappreciation of the crucial role of nature in sustaining our food systems, has contributed to drastic deforestation as well as degradation of land and water systems that in turn affect agricultural productivity.[4] Some 27 percent of global forest loss from 2001 to 2015 can be attributed to land-use change for production of commodities such as soybeans, beef, and palm oil. Other culprits are forestry (26 percent), shifting agriculture (24 percent), and wildfires (23 percent).[5] The near disappearance of some of the world's large rivers and freshwater lakes, such as the Aral Sea and Lake Chad, and the associated public health impacts have been linked to excessive irrigation development.[6] Heavy application of agrochemicals has contributed to the extinction of many insect species, jeopardizing the sustained provisioning of insect-based ecosystem services and threatening a progressive collapse of natural and human-dominated ecosystems alike.[7] Injudicious use of synthetic pesticides, in particular, is associated with biodiversity loss and human health impacts.[8] Agriculture also contributes significantly to greenhouse gas emissions.[9] While intensive, high-input farming that increases land productivity can degrade natural systems, low-input production systems, such as those in large parts of Africa south of the Sahara, can lead to rapid conversion of remaining natural habitats,[10] including tropical forests, but achieve only low productivity levels, which are linked to food insecurity and civil strife.

ENVIRONMENTAL DEGRADATION AND SHOCKS

Among the consequences of habitat loss as agriculture expands is a growing risk of zoonotic and vector-borne diseases. Human incursion into forests and intensification of livestock production can prompt the crossover of pathogens from wildlife to livestock and people, facilitating the spread of zoonotic diseases such as Ebola, SARS, MERS, Lyme disease, and Rift Valley fever. Similarly, expansion of irrigation can increase risks of mosquito-borne diseases including malaria, zika, dengue, and chikungunya. Agricultural drivers were likely associated with more than a quarter of all – and more than half of all zoonotic – infectious diseases that have emerged in humans since 1940.[11] Thus, protecting environmental health is essential to protect human and animal health.[12]

Degradation of ecosystems, together with climate change, threatens the productivity of our food systems; it also causes shocks such as droughts and floods that, like the coronavirus pandemic, affect human livelihoods, health, and long-term options. The biodiversity that underpins food production is disappearing owing to land use changes, pollution, and climate change.[13] Wetlands, tropical forests, croplands, and grazing lands are particularly affected by degradation, increasing risks for food, nutrition, and water security. And climate change affects all these systems, adding another stressor to natural resources, including increasingly frequent natural disasters.

This depletion of vital natural capital often affects the world's poorest people most severely.[14] Land degradation alone already affects the wellbeing of 3.2 billion people.[15] High poverty levels and dependence of livelihoods on unsustainable use of natural resources creates the conditions for a "poverty-environmental degradation" trap – a vicious cycle in which poverty leads to resource degradation, which leads to more poverty, increasing both human vulnerability to natural hazards and the fragility of the ecosystems on which poor people depend.[16]

SYSTEMS THINKING FOR NATURE-POSITIVE FOOD SYSTEMS

By recognizing the dependence of agriculture – and food systems more broadly – on nature, we can focus attention on reducing damage caused by production, or even better, on achieving "nature-positive" production that meets food needs *and* protects and restores ecosystems (Box 1). For example, the ecosystem services provided by noncrop habitat and diverse land use in agricultural landscapes are essential to farming and, if fostered, can reduce dependence on agrochemicals.[17] These services include pest suppression, soil conservation, nutrient retention, and crop pollination.[18] Diverse land uses also support biodiversity and cultural assets essential to human wellbeing.[19] Common agricultural production practices too often degrade ecosystem services, which leads to greater reliance on purchased inputs and an increased propensity to further damage ecosystem services.[20] Breaking out of this vicious cycle requires integrated solutions.

In some cases, innovations in "future foods" can decouple agriculture from nature and reduce environmental impacts. Alternative foods such as mycoproteins, insects, cultured meat, spirulina, sugar kelp, and chlorella hold promise to provide the full spectrum of essential macro- and micronutrients, and could reduce our reliance on traditional production systems for crops, fish, and livestock.[21] But these innovations will not widely replace traditional production in the near future.

BOX 1 ECO-AGRI-FOOD SYSTEMS

"Eco-agri-food systems," a term developed by The Economics of Ecosystems and Biodiversity (TEEB) initiative, emphasizes entire value chains, as well as the ecological, economic, social, and human foundations of food.[a] Natural ecosystems provide some of the largest and most important but economically invisible inputs to most types of agriculture through ecosystem services. Many targets of the Sustainable Development Goals are underpinned by the delivery of ecosystem services, including those pertaining to the food system.[b]

Agricultural ecosystems (agro-ecosystems) are now the world's largest ecosystems. They are actively managed by humans to provide food, fiber, and fuel and draw upon many supporting and regulating services as "inputs" to production, such as soil fertility, nutrient cycling, pollination, and biological pest control. Some of these services, such as soil fertility, are determined at the field or farm scale and, hence, are driven by management choices of private land users; others, such as pest and plant-disease control and pollination, are driven by landscape features including the diversity and spatial configuration of different types of land use. Landscape features are, in turn, determined by varying mixes of private, collective, and public land-rights holders. Some ecosystem services, such as sediment retention and nutrient retention, are delivered at even larger spatial scales (for example, watersheds) and influenced by both exogenous factors such as weather, topography, and management practices adopted by multiple land users.

In sum, a spectrum of local and larger-scale factors typically determines ecosystem service provision. Critically, ecosystem services and agricultural management are closely intertwined – ecosystem services influence management, and management in turn, particularly the choice of techniques, influences ecosystem services. This tight coupling of agricultural and natural ecosystems means that how eco-agri-food systems are managed matters not only to food security and the economy but also to the health of the planet.

Source: [a]TEEB (The Economics of Ecosystems and Biodiversity), *Measuring What Matters in Agriculture and Food Systems: A Synthesis of the Results and Recommendations of TEEB for Agriculture and Food's Scientific and Economic Foundations Report* (Geneva: UN Environment, 2018); [b]S. Wood et al., "Distilling the Role of Ecosystem Services in the Sustainable Development Goals," *Ecosystem Services* 29A (2018): 70–82.

FIGURE 1 Actions for nature-positive food systems

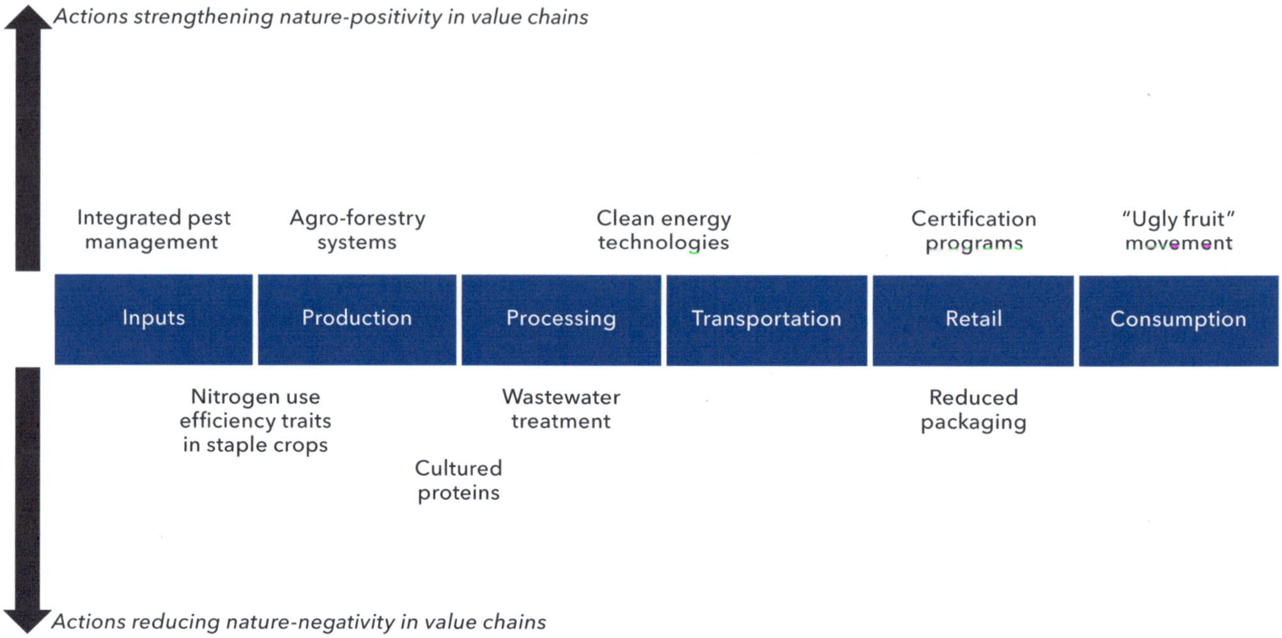

INTEGRATION OF NATURAL AND AGRICULTURAL SYSTEMS

By helping producers use agriculture to restore and maintain ecosystem services, better integration of natural and agricultural systems provides the key to greater sustainability and resilience. For example, sequestration of carbon in agroforestry systems can be increased. Wetlands, if maintained or restored, will provide water storage and purification, buffer climatic extremes, and provide a source of income and food diversification. Habitat conservation and coordinated landscape management at scale can support healthier ecosystems and enhanced biological pest-control services. And expanded collection and conservation of plant genetic resources *in situ* can provide resources for future crop development. Various certification efforts, such as that of the Rainforest Alliance, and – on the consumption side – the "ugly" fruit movement have helped to advance some of these ideas. None, however, is a silver bullet.

Applying systems thinking – that is, taking into account the overall structure, interlinkages, and feedback loops to the food system – can allow us to develop synergistic solutions that support both food security and planetary health. Systems thinking can also be used to redefine food system boundaries, expand the range of externalities that are taken into account, and help us to reconsider what is within the sphere of influence and control of the system; this is important for addressing the interrelated nature of the SDGs that affect food system outcomes.

The proposal to redefine food systems as "eco-agri-food systems," put forward by The Economics of Ecosystems and Biodiversity (TEEB) study, effectively includes all activities along the food value chain in one single system – from ecosystems to farms to intermediaries and processors to retailers and consumers.[22] This allows us to recognize and account for all major externalities along these value chains, and to work toward increasing nature-positive activities and reducing nature-negative activities (Figure 1). Putting this framework into operation will be challenging; nevertheless, the holistic system lens can help identify potential synergies that can contribute to multiple goals.

The linkages between the natural and agricultural systems imply that these systems can improve in tandem if we put in place a virtuous cycle of healthier ecosystems and reduced dependence on external inputs, with further positive implications for the environment.[23] A systems view requires concerted action in many areas, beginning with valuing ecosystem services and investing in them, including appropriate technologies and institutions for different contexts.

Taking a landscape approach is one way to put this into practice. For example, working at the landscape level, sustainable intensification of some areas through more efficient use of water, land, and other inputs, plus adoption of higher-yielding seeds, precision agriculture, and judicious use of (clean-energy-powered) irrigation and fertilizers, followed by solar-powered agro-processing, can reduce pressures on other natural areas, such as forest frontiers and semi-natural habitats, and so prevent degradation and further habitat loss.

To achieve better integration of natural and agricultural systems, appropriate governance will be essential.

GOVERNANCE FOR NATURE-POSITIVE FOOD SYSTEMS

Governance affects how people interact with each other and with the natural resource base. Through a variety of decision-making mechanisms, governance shapes agreements over rights, responsibilities, rules, and regulations, which in turn shape people's behavior and the outcomes for food and natural systems.[24] Although "governance" is often conflated with "governments," the state is not the only source of authority or rules. National laws and government agencies certainly play a role, but more local formal and informal institutions, such as resource user groups, can be equally or more important. Local social norms about respecting (or exploiting) nature also shape behavior. Assessing the extent to which laws, regulations, and the norms of state and local actors value ecosystem services can be an important starting point for understanding the motivations of these actors and identifying effective incentive structures for nature-positive food systems.

GOVERNANCE FOR LANDSCAPES AND COMMUNITIES

Landscape-level governance is particularly important for the sound use of natural resources and provision of ecosystem services. Much agricultural production takes place within a mosaic of land uses, including crop fields, animal grazing areas, forests and agroforestry areas, wetlands, and water bodies. There is also a mosaic of landownership and governance. Crop fields and agroforestry lands are often held privately; grazing lands, forests, wetlands, and water bodies are often held collectively and managed as community commons, even when they are officially under state ownership. Maintaining complex, multifunctional landscapes requires coordination across multiple uses and actors to increase ecosystem services and reduce negative environmental impacts. This generally requires a combination of formal and informal coordination mechanisms. It is challenging for such governance arrangements to keep pace with rapid technological changes, as illustrated by depletion of groundwater aquifers when mechanized pumping allows farmers to extract water faster than it is recharged. Building trust among the actors and sharing reliable information are among the key ingredients for effective deliberation and collective problem-solving.[25]

Multistakeholder platforms (MSPs) are increasingly seen as a useful approach for creating negotiation space and knowledge exchange in integrated landscape initiatives.[26] By bringing together different actors, including government, the private sector, and civil society, MSPs can help achieve a shared understanding of how each actor's behavior affects others, and facilitate agreements on rules, rewards, and penalties for actions that increase or decrease ecosystem services. The Nairobi Water Fund, for example, is a recently formed MSP that aims to improve the efficiency and allocation of water use across the Upper Tana watershed (Box 2). However, MSP arrangements are not a panacea: power differences among groups will affect whose voices are heard, and some agreements reached may not be enforceable. Government actors can lend legitimacy to MSP decisions and support follow-up on agreements, thus enhancing trust and enforcement of agreements.

INCENTIVES FOR SUSTAINABILITY

Policies for sustainable management of resources have largely focused on regulatory instruments that decree what people may or may not do, such as banning burning or cutting down trees, in order to reduce "nature-negative" impacts. However, a number of policies reward nature-positive practices that sustain environmental services. For example, payment for ecosystem services (PES) initiatives provide financial rewards for sustainable practices, such as paying people to plant or keep trees.[27] Watershed management programs designed to manage upstream-downstream linkages of water, soils, and pollutants likewise offer

BOX 2 NAIROBI WATER FUND ENCOURAGES SUSTAINABLE PRACTICES IN THE UPPER TANA RIVER WATERSHED

The Nairobi Water Fund was recently set up by The Nature Conservancy and partners to improve downstream water quality for millions of water users and provide benefits to farmers from more sustainable water use. The Fund provides farmers in Kenya's Upper Tana River Basin with support for small-scale water-storage development, grass and tree seedlings, and expert advice on sustainable agricultural practices, including terracing, planting napier grass for erosion control and livestock fodder, and maintaining vegetative buffers near the river's tributaries. For farmers, these practices have potential to increase crop yields. Downstream benefits include reduced water-treatment costs and increased hydropower generation. Together, these benefits are sufficient to create a "business case" for public and private sector funding. Engagement of various stakeholders was important in creating buy-in.[a] However, even with these engagement processes, women and farmers with small landholdings are often unable to adopt space-consuming sustainable land management practices, such as use of water-storage pans or setting land aside for buffers; identifying the preferences and abilities of both women and men could help address this disparity.[b]

Source: [a]A.L. Vogl et al., "Valuing Investments in Sustainable Land Management in the Upper Tana River Basin, Kenya," *Journal of Environmental Management* 195, Part 1 (2017): 78–91. [b]R. Nijbroek and E. Wangui, *What Women and Men Want: Considering Gender for Successful, Sustainable Land Management Programs: Lessons Learned from the Nairobi Water Fund*, GLF Brief 7 (Global Landscapes Forum, CIFOR, 2018).

payments in cash or kind to upstream farmers and residents to change their practices in ways that improve the flow and quality of downstream water.

Both coercive regulatory and remunerative payment approaches require considerable information for implementation, enabling institutions (such as tenure security), and enforcement authority (Chapter 2). This makes it challenging for state agencies or PES schemes to operate locally appropriate programs over large areas. However, state efforts can be complemented by local institutions and by normative incentives. Norms regarding the environment can strongly shape people's behavior, especially in cooperating for collective goods.[28] Norms for sustainable practices can be strengthened by changing the mindset of officials and community actors through campaigns that highlight the importance of ecosystem services, and that recognize and praise certain behaviors (such as planting trees) or create a stigma for other behaviors (such as deforestation).

Given the urgency and complexity of transforming food systems to meet the needs of all, both today and in the future, all three types of incentive mechanisms are needed: regulation, financial rewards, and "nature-positive" norms. Experience has shown that sustainable solutions require not only technical information and government intervention, but also active stakeholder involvement through a variety of participatory processes that tap into local knowledge and objectives.[29] Box 3 illustrates this through the experience of the Promise of Commons initiative in India.

NATURE-POSITIVE APPROACHES

Food systems played a key role in the enormous improvements in living conditions registered over the past 50 years. Yet, many food systems are now widely viewed as seriously flawed or "broken";[30] nearly 10 percent of the world's population suffers from persistent hunger; low-quality diets for many perpetuate malnutrition over generations; and the environmental costs of our food systems affect climate, water, and food production. COVID-19 has been a wake-up call for agriculture and food systems to proactively and more aggressively preserve natural systems to reduce the risk of future disease outbreaks, natural disasters, and climate change that may cripple national economies and people's livelihoods.

BOX 3 PROMISE OF COMMONS IN INDIA

In India, over 80 million hectares are classified as "commons," including community forests, pastures, and water bodies. These common areas provide a source of livelihoods for over 350 million people through fodder and forest products as well as critical ecosystem services, including provisioning of surface and groundwater, soil nutrients, pollination, and pest control. Yet many of these commons have experienced encroachment, decreased vegetative cover, and declining water availability.

To counter these problems, the Foundation for Ecological Security (FES) is using a systems approach to develop the multi-organizational Promise of Commons Initiative. The core elements of this approach are securing community rights over the commons; strengthening collective action for responsible local governance of resources; and access to resources and finance for restoration activities. Securing collective rights is seen as essential to creating the incentives and authority for managing common resources. In addition to forming or strengthening local organizations such as forest- and water-user groups both directly and in partnership with other organizations, FES also pools technical tools, such as water budget calculators, and provides them to communities to help them plan where to plant trees or place water-harvesting structures so as to optimize ecosystem services at the landscape level. The initiative then helps communities tap into rural employment program funds or other public funds to finance restoration projects.

Together with 83 NGO partners, FES currently works with 29,221 villages across 10 ecological regions of India, restoring 8.71 million acres of common lands. FES also works with government agents at the state, district, and subdistrict levels to promote recognition of the value of the commons and overcome the notion that these are "wastelands." FES convenes multistakeholder platforms that bring government and other agencies together with communities to identify needs and opportunities to improve both governance and technical interventions in the commons in ways that also benefit private agricultural lands and livestock production.

Evaluation studies show marked improvements in a range of ecological and economic indicators related to the Promise of Commons. Standing biomass on the community-managed lands is 2.6 times greater than on unmanaged common lands; and the grass biomass is 2.4 times greater.[a] Secure rights and better resource conditions have improved the availability of water, fodder, and nontimber forest products, leading to income gains. On average, household incomes have increased by 30 to 50 percent over a period of 3 to 5 years, and the value of benefits derived from the commons per household per year increased from US$130 to $330.[b]

Increased incomes and savings have allowed the majority of households to increase spending on food, education, health, housing construction and repair, and purchase of agricultural inputs and assets. The estimated value of returns, including environmental benefits, are eight times the cost of the investment.[c] A recent comparison between communities where FES had been working and comparable communities where they had not intervened shows that the coronavirus pandemic had negative effects on farm and nonfarm livelihoods in both types of communities; but those where FES had been working were significantly less likely to resort to risky coping strategies, such as eating seed stocks or taking out loans. This suggests that the FES interventions aimed at improving livelihoods and ecological health have also strengthened resilience.

Source: [a]FES (Foundation for Ecological Security), *Ecological Health Monitoring: A Summary Report 2016–17*, Anand, India, n.d.; [b]FES, *Restoration of Ecological Security as a Means of Improving and Securing Livelihoods in the States of Karnataka and Rajasthan: A Mid-term Assessment*, NR Management Consultants, New Delhi, n.d.; [c] D. Mondal, S. Singh, and J.V. Dhameliya, "Assessing the Value of Our Forests: Quantification and Valuation of Revegetation Efforts," FES Working Paper (FES, 2005).

While technological solutions, such as cultured meat, offer promise, the largest potential for gains lies in better integration of natural and food systems. Realizing these gains requires broader perspectives that integrate complete landscapes and watersheds and consider these as integral to our human systems. These broad integrative perspectives differ from common reductionist approaches that focus on fields or farms. Nature-positive approaches require new forms of governance that give individual actors confidence that the ecosystems services on which their operations depend will also be nurtured by other actors in the system.

These governance issues often present the thorniest challenge to improving natural and food system integration. Polycentric governance – separate institutions with their own roles and responsibilities, but that are working toward a common objective – presents a promising path forward. Given the scope of the challenges, multiple-incentive mechanisms, such as regulations plus subsidies, can be usefully employed by engaged institutions. At the same time, the potential role of nature-positive norms must not be forgotten as communities of all sizes grapple with the common property issues that almost invariably arise at the interface of natural and food systems. This is particularly true in lower-income settings, where externally imposed regulations are difficult to enforce, subsidies are difficult to finance, and penalties would have excessive negative impacts on those who are already poor.

The integral dependence of food systems on natural systems means that food systems must work with, and not against, natural systems. Achieving better integration between natural and food systems can be eased by technological solutions; however, governance issues across eco-agri-food systems will likely be more fundamental to transformation and, as such, should be a focus of continued research and experimental practice. Only then will we be able to more proactively address the growing risk of zoonotic diseases and other risks from environmental degradation and disasters that threaten to disrupt our food systems and our lives.

> "With greater investment in the integration of natural systems with human systems, we can move toward 'nature-positive' systems."

CHAPTER 5

Toward Inclusive Food Systems
Pandemics, Vulnerable Groups, and the Role of Social Protection

NEHA KUMAR, AGNES QUISUMBING, AULO GELLI, UGO GENTILINI, AND SARA SHAPLEIGH

Neha Kumar, **Agnes Quisumbing**, and **Aulo Gelli** are senior research fellows, and **Sara Shapleigh** is a program coordinator, all in the Poverty, Health, and Nutrition Division, International Food Policy Research Institute, Washington, DC. **Ugo Gentilini** is a global lead for social assistance, World Bank, Washington, DC.

KEY MESSAGES

- Poverty, poor health, and malnutrition not only result from pandemic stresses, but they also contribute to pandemic-related risks that impact wellbeing and worsen existing inequities.
- Vulnerable groups have been most affected by disruptions to food systems, such as lockdowns, through loss of employment and incomes.
- The urban poor, especially informal workers and women, have borne the brunt of health and employment impacts. Refugees and internally displaced persons have also been disproportionately affected.
- Men, women, and children experience different risks and stresses. Women have been more likely to experience increased domestic violence and food insecurity, reduced autonomy, and loss of income.
- Social protection is critical for supporting vulnerable groups and has expanded to an unprecedented degree. But many people were still left without coverage, and programs were rarely gender sensitive.

RECOMMENDATIONS

- Develop a better understanding of pandemic-related risks and stresses as well as requirements for effective social protection and related financing during a crisis.
- Use evidence to support policymakers' efforts to strengthen food system resilience to support vulnerable groups. Test and document local, context-specific innovations.
- Protect, rebuild, and strengthen women's rights and control over assets, both during the pandemic and for long-term food system transformation.
- Strengthen women's ability to build social capital by supporting group-based programs, which can also be effective service-delivery mechanisms.
- Adopt and improve on innovative responses to the pandemic, including use of digital transfers and expansion of social protection to informal workers, with attention to reducing inequities such as the digital divide.

- Consider complementary programming on gender-based violence, mental health, and maternal and reproductive health for all social protection programs.
- Distill lessons from pandemic responses so that social programs and policies can be redesigned to reduce gender, ethnic, and other inequalities to limit the impact of the next pandemic and promote an inclusive food system transformation.

Across our food systems, there is an urgent need to understand the effects of the current pandemic, including the impact of policy responses such as the stay-at-home orders and social protection systems, on the wellbeing and nutrition of vulnerable populations. Pandemics can be expected to have a disproportionate impact on minorities and other disadvantaged groups – who often suffer more from preexisting conditions and face greater risk of infection, and whose coping options are constrained by poverty and limited access to services, information, and technology.[1]

Emerging data on the impacts of the coronavirus pandemic have exposed underlying inequalities. Importantly, poverty and poor health and nutrition can both result from pandemic stresses and contribute to pandemic-related risks. The pandemic and related policy responses are affecting the wellbeing of vulnerable populations in part through loss of access to healthy diets; in turn, malnutrition may raise the risks of serious infection, complications, and death. Moreover, the actions some households take to cope with the short-term impacts of the shocks (such as selling assets and taking children out of school) can affect their livelihoods and wellbeing in the long run. These immediate and secondary effects and their feedback on food systems, nutrition, and health form a complex web of trade-offs and impacts.

To better understand these impacts and potential solutions, we examine the problem from two perspectives. Viewing the problem using a COVID-19 lens reveals both economic and health inequalities and vulnerabilities. Viewing the problem through a gender and inclusion lens shows how the pandemic magnifies these underlying inequalities and vulnerabilities. When we acknowledge these inequalities and address them in the design of policies, particularly social protection interventions, vulnerable groups get a better chance at attaining a healthier and more productive life – not only during the pandemic, but over the long term as we work toward transformation of our food systems.

DID COVID-19 INCREASE INEQUALITY?

We are just beginning to understand how, and why, this global shock has affected different groups within countries and different members within each household. Here, we focus on how pandemic-related disruptions to food supply chains in developing countries are affecting the poor and other vulnerable groups. The impacts are expected to be widespread but uneven, with rural areas insulated from the worst of the crisis and urban populations harder hit by loss of incomes and employment in the short term (see Box 1 on Myanmar) and medium term.[2] Disruptions in food supply chains can also have implications for diet quality (Chapter 3), with differences between rural and urban areas, as seen in the case of Myanmar. Rural areas may be better protected from decreased food and dietary diversity, at least in the short run, as seen in Burkina Faso (Box 2).

URBAN AND RURAL IMPACTS

Urban areas are the epicenter of the pandemic, with over 90 percent of all COVID-19 cases reported.[3] A recent review summarizing the impacts of COVID-19 on cities emphasized the disproportionate impacts on poor, marginalized, and vulnerable groups.[4] These differential health impacts reflect the urban poor's limited access to medical care and their substandard living and working conditions. Social distancing and sanitation – key preventive measures for a disease like COVID-19 – are difficult to achieve in urban slums.

In terms of employment, the pandemic lockdowns have disproportionately affected workers in the informal sector. This reflects the nature of food supply chains in much of Africa and Asia where the midstream and downstream services, which are labor intensive and face little regulation or public standards, are concentrated in dense urban and peri-urban areas. Informal workers generally have neither employment nor health insurance. Yet, the informal sector may prove nimbler than the formal sector in coping with the pandemic. In India, for example, informal shops (relying on family labor) have been active during times when formal retailers had to close because of the pandemic.[5] In Africa, in contrast, nonpoor urban households have been hit significantly as manufacturing and businesses face the strictest lockdowns.[6]

Social safety nets in urban areas may also help to buffer households and individuals against shocks, although coverage is generally quite low (Box 1); in low-income countries, only 18 percent of the poorest fifth of the population was covered by a social assistance scheme before the pandemic.[7] To address the multiple crises affecting health, equity, employment, public finance, and public services in urban areas, governments must simultaneously help those who are infected and also address the urgent need to curtail the spread of the disease and to use safety nets to mitigate the food security impacts stemming from loss in incomes and employment.

In **rural areas**, dominated by small farms dependent on family labor, the farm sector has not suffered major direct impacts but may be indirectly affected by disruption of input supply chains and a dampening of demand for food due to reduced incomes. Even within the farm sector, the impacts may be uneven and are likely to vary across countries and contexts. Work on Ethiopia's vegetable chains suggests that medium-scale tenant farmers have been most affected by labor disruptions (Chapter 6).[8] Small farms that rely on family labor and large farms that hire more labor but are big enough to reorganize labor use safely have done relatively better. At the same time, medium-scale farmers benefited more from the price increases in output markets.

BOX 1 A RURAL-URBAN COMPARISON FROM MYANMAR

In June and July 2020, approximately 2,000 mothers from nutritionally vulnerable groups in Yangon, Myanmar's largest city, and rural agricultural areas of Myanmar's dry zone were surveyed. The urban sample was comprised of women who were pregnant in January 2020. Survey questions about the perceived impact of the pandemic on the women's households revealed some interesting differences between the rural and urban samples. While loss of income or jobs was the most frequently reported impact across both samples, a larger share of urban households reported this setback. Among households able to estimate their monthly incomes confidently, incomes declined between January and June 2020; and a larger share of rural households than urban households reported an income decline or zero income as of June 2020. There was also a sharp increase in income-based poverty – up 31.9 percentage points among the rural sample and 21.9 percentage points among urban households. Two factors were likely to increase the risk of becoming poor: job loss of a family member (including migrants) and the birth of a child (but not pregnancy), indicating a lack of access to social protection and maternity benefits at the workplace. Coping mechanisms did not differ much between the urban and rural samples, with many households taking out loans, drawing down savings, or simply reducing spending.

Urban women were twice as likely as the rural respondents to report food supply challenges (related to availability, prices, or affordability) as a result of the pandemic, and fared worse on every indicator of food security, including diet quality. About 34 percent of the urban sample reported inadequate maternal dietary diversity as compared with about 16 percent of the rural sample. Plausible explanations for the difference in diets include better access to more nutritious foods in rural areas at this time of the year, food supply disruptions in urban areas, and food taboos related to pregnancy/post-pregnancy (since the women in the urban sample were either pregnant or had recently given birth). The pandemic also had a major impact on mothers' diets, particularly increasing the risk of micronutrient deficiencies.

Myanmar's government acted quickly and implemented its COVID-19 Emergency Response Plan in April 2020, which included in-kind food transfers to vulnerable populations; emergency food rations delivered via community-based food banks; top-up benefits for existing maternal, child, and pension program recipients; and cash transfers to the most vulnerable households. Approximately 30 percent of the respondents reported receiving cash or in-kind transfers in June 2020, primarily from the government. Coverage of the transfers was higher in the rural sample and among households with fewer assets. However, within the asset-poor group, only 35 percent received transfers – indicating a significant gap in coverage.

This experience in the early days of the pandemic points to the need to scale up social protection efforts to include urban populations, to expand coverage of pregnant women and mothers of young children, and to adjust labor laws to provide protection to women's incomes during pregnancy, childbirth, and early infancy.

Source: D. Headey et al., "Poverty and Food Insecurity during COVID-19: Evidence from the COVID-19 Rural and Urban Food Security Survey (RUFSS) – June and July Round," *Myanmar SSP Policy Note* (Washington, DC: IFPRI, 2020).

VULNERABLE GROUPS

Informal sector workers live hand-to-mouth with minimal savings, and must work every day to feed themselves and their families. About 61 percent of the world's employed population is in the informal sector, and 93 percent of the world's informal employment is located in emerging and developing countries.[9] Unlike many jobs in the formal sector, most informal sector jobs cannot be performed remotely,and thus any restrictions on movement or lack of protective equipment fall more heavily on informal sector workers.[10] The pervasiveness of informality in rural areas also implies that much of the world's rural population may not be protected from loss of employment. This highlights the need for social safety nets beyond those that are targeted to the poorest of the poor.

Migrant workers were especially vulnerable during national lockdowns. Many were forced to return home, often without any means to do so or safety nets to fall

BOX 2 IMPACTS OF THE PANDEMIC ON DIETS, NUTRITION STATUS, AND FOOD SECURITY IN RURAL BURKINA FASO

In Burkina Faso, an observational study aimed at providing new empirical evidence on trends in diets, health, nutrition, and food security among rural populations was temporarily interrupted by the pandemic, allowing for analysis of the impact of the crisis. Three rounds of surveys had been conducted in 90 villages in three regions of the country between early 2017 and mid-2019, including approximately 1,080 households with mother-and-child pairs. The fourth survey in the series, which began in March 2020, was interrupted by the national lockdown in April and resumed in June after lockdown measures were eased. The study looked at child nutrition (body mass index [BMI]), child morbidity (caregiver-reported symptoms during previous week), diets (for women and children, using the measure of micronutrient intake), and household food security (shocks and access to food and to social assistance).

The analysis found that mean BMI was 0.24 points higher in children surveyed post-lockdown compared to children surveyed pre-lockdown. No statistically significant differences were found in children's and mother's diets, or in household food insecurity, shocks, or social protection. However, the small change in social assistance scores may reflect the scale-up of some targeted social assistance post-lockdown. Though no evidence was found of a major food-security-related shock in the immediate aftermath of the April 2020 lockdown, the data suggest that these rural households were already struggling with high levels of food insecurity well before 2020. Moreover, the data also suggest these households were rarely covered by social assistance programs either before or after the lockdown, highlighting an important gap in safety net coverage.

Source: A. Gelli et al., "Diets, Nutritional Status, and Food Security in Rural Burkina Faso: Empirical Evidence Before and After COVID-19 Lockdown" (forthcoming).

back on.[11] The chaos caused by the sudden national lockdown in India is now infamous – millions of migrant workers made long treks back home on foot, as all transportation was shut down, and many died during their journey.[12] In countries where safety nets were already in place or where a database of workers existed, it was easier to provide assistance to such workers (Box 3).[13]

Refugees and internally displaced persons (IDPs) in refugee camps and temporary settlements have been affected disproportionately by the pandemic. Like slum dwellers, they face overcrowding and lack access to water, hygiene, and sanitation. Making matters worse, many refugee and IDP populations are located in places with weak or fragile healthcare systems and low government capacity. Despite this risk, the global response to the pandemic has not adequately included refugees and IDPs.[14] A study that modeled COVID-19 transmission rates in Rohingya refugee camps in Bangladesh shows that dealing with COVID-19 would require large increases in healthcare capacity and infrastructure that may exceed what is feasible.[15] This highlights the need to think and plan ahead for these settings and for the next pandemic. In the longer term, this will require not only investments to make the camps more livable, but also services to move people out of camps more quickly and into more sustainable and permanent settings so that such crises can be avoided.

GENDER AND FAMILIES

Pandemic stresses on vulnerable populations – including financial stresses and confinement – can create or aggravate tensions in the home, lead to mental distress, and reduce women's autonomy. For **men**, who are typically seen as and consider themselves to be their families' breadwinners, loss of employment and income may lead to depression, suicidal thoughts, or domestic violence.[16] Men may also lose contact with their peers, exacerbating stress.[17] As male migrants return home, women are likely to lose their autonomy in decision-making and their role as de facto household heads.

BOX 3 DEVISING A SAFETY NET DURING A PANDEMIC: A CASE FROM BIHAR, INDIA

In the wake of India's pandemic-induced lockdowns, the state government of Bihar implemented the Corona Sahayata program. This social safety net was set up from scratch in the middle of the pandemic to support migrant workers from Bihar who were stuck in other parts of the country and had no access to safety nets. Given the circumstances, it was implemented and executed completely digitally. The program was far from perfect but is an excellent example of what can be accomplished quickly in a crisis.

In late March 2020, India's federal government announced a countrywide lockdown and soon after, a relief package providing cash and food transfers. India's Direct Benefit Transfer platform is used by various federal and state programs to deliver transfers digitally. This platform is complemented by the biometric ID system (Aadhaar) and a nationwide financial inclusion program (Jan Dhan) designed to expand bank account ownership. The pandemic provided the perfect opportunity to test these systems. It soon became clear that the government's biggest problem was not how to pay beneficiaries but whom to pay. While India has a multitude of social safety nets – at both the federal and state levels – it does not have a unified beneficiary database or social registry.

Many migrant workers from Bihar were stranded in other states, with no income and no access to social assistance. Social assistance was available to them only in Bihar, where they were registered. On April 7, 2020, the Bihar government responded by launching the Corona Sahayata scheme to provide direct cash transfers of 1,000 rupees (about US$15) into bank accounts of migrant workers, so that they could return home. Starting from a blank slate during a pandemic meant that the scheme had to be digital end-to-end. Applicants had to provide proof of residence outside Bihar and a bank account registered in Bihar. They also needed access to a mobile phone to receive the transfer. The Bihar government received about 3 million applications within a month of the scheme's launch, and roughly 2 million were verified and paid before the end of May 2020.

Not surprisingly, there were challenges in implementation, particularly the application process and complicated eligibility criteria, and the costs often outweighed the benefits. There are no data to evaluate how the scheme performed, but the experience offers important lessons for safety nets in such circumstances. Even in a country where the prerequisites of digital transfers were up and running, identifying and paying beneficiaries was not straightforward. With restrictions on mobility, bank accounts may not be the best option for delivering transfers. In addition, as is the case with all digital transfers, people who lack digital access are excluded.

Source: A. Mukherjee, "Digital Cash Transfers for Stranded Migrants: Lessons from Bihar's COVID-19 Assistance Program," *CGD Notes* (Washington, DC: Center for Global Development, 2020).

For **women** and **children**, quarantine conditions increase intrahousehold tension and exposure to potential violence. Intimate partner violence has been shown to increase during pandemics (Box 4), but overburdened health services – often the first point of contact for women seeking help – may be unable to respond. Redirection of health services toward the pandemic response also jeopardizes reproductive health services such as family planning and obstetric care. The indirect effects of an inadequate health-care system will also have long-lasting consequences for maternal and child health – as the COVID-19 response takes precedence over non-emergency health services, such as growth monitoring and nutritional counseling.[18]

Stay-at-home orders also make it difficult for many women to procure food for cooking. Women may have to allocate the limited amount of time permitted outside the home to procure either safe water or food for their children and families; the difficulty of accessing water or food is a long-term problem that is magnified by the pandemic. And food insecurity may affect women more than men, as seen in previous work on the food price crisis of 2007–2008.[19] As their small businesses collapse and their informal work arrangements are canceled, women lose financial independence, affecting their empowerment in the short term, with potential longer-term impacts on children. Children may be affected in multiple ways – through increased exposure to domestic violence, poorer nutrition and

BOX 4 PANDEMICS AND VIOLENCE AGAINST WOMEN AND CHILDREN

Early in the pandemic, stakeholders working on gender equality warned of an increasing risk of violence against women and children due to a confluence of factors, arising from both direct effects of the virus and associated response measures. A review of the literature on past pandemics, public health emergencies, and economic crises identified nine distinct and overlapping pathways through which pandemics could lead to changes in the frequency and severity of violence, depending on the setting and population group of interest. Primary pathways include (1) increases in economic insecurity and poverty-related stress, and (2) social isolation and quarantines, which in turn induce poor mental health and increase exposure to perpetrators in the home. Emerging literature has largely confirmed these dynamics.

More than 70 rigorous qualitative and quantitative studies have been published since the onset of the pandemic, most of which attempt to measure trends in violence over the pandemic period or to identify associated risk factors. While the majority of these studies find evidence of increased incidence and/or severity of violence, there are clear data constraints, including the use of reported data from administrative sources (which may also be affected by the pandemic), that likely affect the findings. Moreover, most of the evidence is from high-income countries and focuses on intimate partner violence, rather than more diverse forms of violence or geographic locations. Although the evidence with respect to effective response options is scant, the broader literature suggests that, in addition to bolstering first-response efforts, it is critical that social protection be put in place to address economic factors along with policies ensuring safe housing for survivors. Long-term solutions include increasing funding for violence-related services, integrating violence risk into pandemic preparedness planning, and changing social, legal, and economic policies that condone and allow violence against women and children to continue.

Source: A. Peterman et al., "Pandemics and Violence against Women and Children," CGD Working Paper 528 (Center for Global Development, Washington, DC, 2020); and A. Peterman and M. O'Donnell, "COVID-19 and Violence against Women and Children: A Third Research Round Up for the 16 Days of Activism," *CGD Notes* (Washington, DC: Center for Global Development, 2020).

health, lower schooling attainment, and developmental issues owing to all of the above. These impacts may hurt girls more, which in turn could affect female labor-force participation in the next generation.

Strengthening women's assets should be a key priority both in pandemic response and recovery and in the longer term. Because women's assets are often the first sold in economic crises,[20] protecting them to the extent feasible and rebuilding them after the pandemic will be crucial. Even more important will be securing women's rights to and control over their assets in the longer term, so that these assets will not be vulnerable to depletion if there is a shock. Such efforts also support women's empowerment. Owing to social distancing and lockdowns during the pandemic, many women have lost access to group-based programs that had provided a way to build both social and financial capital. Rebuilding the social capital embedded in women's groups may also empower women to learn about and avail themselves of public services, and to provide the leadership their communities need.[21]

HOW INCLUSIVE WAS THE SOCIAL PROTECTION RESPONSE TO COVID-19?

Social protection is one of the key measures that governments take to reach vulnerable people in response to crises. The social protection response to the pandemic has in many ways been unprecedented in scale and nature. About one-third of the 1,414 social protection measures taken by 212 countries and territories, as of mid-December 2020, encompassed various forms of cash transfers. These transfers have reached more than 1.1 billion people, or 14 percent of the world's population – a rate that increases to 16 percent if all forms of social assistance are considered. Relative to pre-pandemic levels, cash-transfer benefit amounts doubled, and coverage (the number of beneficiaries) grew by 240 percent. At least US$800 billion has been invested in social protection, a level 22 percent higher than in the recession of 2008–2009. Programs have been simplified both in administration and design, allowing for faster scaling up during this crisis.[22]

Yet the social protection responses were limited in many ways: cash transfer programs lasted 3.3 months on average, with a mere 7 percent of them extended for a longer period; 30 percent of programs were one-off payments. Additionally, scale-up occurred better and faster where delivery mechanisms were stronger (Chapter 2). Only a quarter of countries reached more than a third of their population, with no low-income country reaching this coverage level. Spending amounted to about US$6 per capita on average in low-income countries, almost 90 times lower than average benefits in high-income countries.[23]

Against this backdrop, governments have leveraged the pandemic response to address some longstanding challenges, many of which can lead to long-term transformation in the design of these programs.[24] For example, at least 11 countries in Africa have extended coverage of cash transfers to vulnerable urban dwellers, including adapting design and delivery practices to fit urban contexts.[25] In India, transfers have been provided to migrants,[26] while a number of countries (such as Thailand and Brazil) are now deliberately supporting informal sector workers by allowing them to apply to receive transfers; this is leading to "quasi-universal" social protection provisions.[27] In some countries, innovative combinations of social assistance and insurance have been introduced (for example, in Pakistan), and interventions like public works are being reconfigured to fit the nature of the pandemic – for example, by waiving work requirements (in India and Ethiopia), changing implementation so workers are not in close proximity to each other (in Uzbekistan), and adapting "cash for services," such as sanitization of homes and neighborhoods (in the Philippines) and remote mentoring of adolescent girls (in Uganda).[28]

Despite these innovations, some opportunities have been missed. Thus far, the social protection response to the pandemic in most countries has not been gender-sensitive, with fewer than 30 programs across 25 countries including gender-sensitive components.[29] This represents a mere 2 percent of all measures that were undertaken across the 212 countries. While this is not surprising given the speed with which governments had to respond, it is worrying because the pandemic-induced crisis can be expected to widen many gender inequities. From a long-term transformation perspective, it is important that policy actions be evaluated from a gendered perspective to ensure, to the extent possible, that the responses are gender-sensitive and built into programs in the future. Complementary programming focused on gender-based violence, mental health, and maternal and reproductive health should all be considered in the design of social protection programs – not just in response to shocks but also to foster long-term change that can prevent financial and health-related disruptions.[30] For example, mobile transfers directly to women's accounts are a promising innovation; however, we need to be mindful of women's relatively limited access to digital technologies.

There is no silver bullet for the diverse and multifaceted problems arising from a pandemic or similar shocks, but local, context-specific innovations may provide a way forward (Box 3). Testing and documentation of such innovations, including the contexts in which they function, coupled with the political will to scale up strategies that work, may provide solutions now and in the future. For example, in countries with limited Internet connectivity and inequality in the distribution of digital devices, the digitally marginalized are not only cut off from important sources of health and economic information, but also deprived of the possibility of benefiting from rapid, targeted safety-net responses such as that implemented in Bihar. Thus safety nets alone are not enough.

PREPARING FOR THE NEXT PANDEMIC AND BEYOND

The coronavirus pandemic is not yet over, and it is unlikely to be the last pandemic we will face. Fully understanding the short-, medium-, and long-term implications of the pandemic and policy responses across the food system will require integrated, multidisciplinary efforts by the policy, program, and research communities.

The COVID-19 experience makes clear who are the most vulnerable groups – those who live in suboptimal conditions, depend on unreliable livelihoods, and do not have access to healthcare or safety nets. Digital technologies have shown promise in the current crisis but have also highlighted the challenges of the great digital divide between rich and poor and of covering individuals and families that are off the grid. Investments are needed not only in digital technologies and infrastructure but also in generating a database of potential beneficiaries that can then be easily linked to the digital infrastructure.

Many governments responded quickly to the crisis, but the response was often insufficient. To enable a sustainable and proactive approach, a better understanding is needed of the requisite level and frequency of safety-net transfers and of mechanisms for financing them. Just as the scientific community has learned much about preventing, mitigating, and eventually stopping the spread of the pandemic, the policy community has also learned lessons about how to mitigate the adverse economic effects, particularly for vulnerable groups. It will be important to distill the lessons learned from the pandemic responses all over the world, so that food systems and social programs and policies can be redesigned to alleviate existing inequalities that hamper the ability to cope with, and recover from, these health crises and to allow everyone to prosper.

The unprecedented nature of the crisis led to innovative responses, paving the way forward in three ways.[31] First, governments responded and reached a large majority of vulnerable households despite various barriers – showing that reaching these households is above all a matter of political choice. Second, efforts must be deliberate and intentional, with a clear goal. Governments are undoubtedly central to these efforts, but it is equally important to recognize that communities, the private sector, and partnerships among them can contribute to achieving long-term objectives. Third, the rapid mobilization of research, both on ways to fight the virus and to understand and mitigate its effects, has increased knowledge on how we might cope with future shocks and enable long-term transformation in social programs and policies.

Preparedness is important, and researchers can play their part in helping policymakers be more proactive and innovative in their endeavors to strengthen food systems. Given the evidence that the pandemic has had disproportionately adverse impacts on vulnerable populations, we must pay attention to health equity, gender and ethnic equality, and human rights in this transformational journey.

"When we acknowledge inequalities and address them in the design of policies, particularly social protection interventions, vulnerable groups get a better chance at attaining a healthier and more productive life – not only during the pandemic, but over the long term."

CHAPTER 6

Food Supply Chains
Business Resilience, Innovation, and Adaptation

THOMAS REARDON AND ROB VOS

Thomas Reardon is a professor in the Department of Agricultural, Food, and Resource Economics at Michigan State University, East Lansing. **Rob Vos** is director of the Markets, Trade, and Institutions Division of the International Food Policy Research Institute, Washington, DC.

KEY MESSAGES

- The pandemic disrupted food supply chains through government-imposed lockdowns and restrictions, affecting labor supply, input provisioning, logistics, and distribution channels, and shifting consumer demand for food.
- Impacts differed by the degree of integration and modernization of food supply chains.
- "Transitioning" supply chains were the most vulnerable – these chains are long but still poorly integrated, face infrastructure limitations, and are dominated by small and medium enterprises (SMEs) that depend heavily on hired labor.
- Traditional supply chains also suffered but less so, being generally short and relying on family labor.
- Modern, integrated supply chains were better positioned to adapt and innovate. Businesses that were able to "pivot" or innovate rapidly fared well, using either their own capacity or intermediaries to expand e-platforms for supply and delivery.
- Ongoing trends, most notably the growth of supermarket-style retail, e-commerce, and food delivery, were accelerated by the pandemic.
- Recent innovations offer opportunities for SMEs in food supply chains. E-commerce jumped by 100 percent in middle-income countries, and specialized logistics intermediaries have innovated to meet the needs of both large food businesses and developing-country SMEs and smallholder farms.

RECOMMENDATIONS

- Create a business environment that supports private sector firms in their central role in food system resilience and transformation.
- Promote food system modernization and innovations – driven by the private sector but enabled by the public sector – that enhance resilience and help generate employment and better livelihoods along food supply chains.
- Develop regulations that promote market integration and reduce transaction costs along supply chains.
- Invest in adequate basic infrastructure, ICT, and mobile networks to facilitate business and supply chain innovation and modernization, especially for SMEs.
- Focus government interventions on targeted support to improve market access and entry for SMEs.

The COVID-19 pandemic has revealed both the vulnerability and resilience of food supply chains. From farm to retail, supply chains have been disrupted, primarily by government-imposed lockdowns and other restrictions affecting labor supply, input provisioning, logistics, and distribution channels. (See Figure 1 for a simplified presentation of food supply chains.) Impacts have differed by type of commodity: Mechanized production of cereals and other staples on large farms proved less vulnerable than labor-intensive and labor-dense production of fruits and vegetables on smaller farms. Postharvest handling, packaging, and processing for many perishable foods were susceptible to outbreaks of COVID-19 among workers because of close working conditions. Disruptions and enhanced virus transmission were especially notable in the meatpacking industry.[1]

Impacts have also differed by country and degree of integration and modernization of food markets. Countries where food systems are transitioning from traditional to modern appear to have been most vulnerable to supply disruptions and restrictions on labor movements. Transitioning supply chains are long and operations often depend on hired labor, but the multiple stages between farm and retail are still poorly integrated and fragmented – characterized by, for instance, little development of temperature-controlled storage and transportation, poorly connected service and input markets, and underfinanced suppliers. These supply chains have been vulnerable to COVID-19 restrictions. Border closures and curfews have led to food losses as transport of perishables have had to take place in daytime heat instead of during cooler nights. In other cases, hired workers were unable to report to work.[2] More traditional supply chains (depicted in Figure 1) have also proven vulnerable for much the same reasons, but less so, as these chains remain short and producers and operators are mostly family owned, using little hired labor.

This chapter draws on collaborative research by the authors with A. Heiman, L. Lu, C. Nuthalapati, and D. Zilberman ("'Pivoting' by Food Industry Firms to Cope with COVID-19 in Developing Regions: E-commerce and 'Co-pivoting' Intermediaries," *Agricultural Economics* [2021, forthcoming]); the case studies published in a special issue of *Agricultural Economics* on "COVID-19 Impacts on Global Food Systems and Household Welfare," ed. J. Swinnen and R. Vos (2021, forthcoming); and the framework provided in T. Reardon and J. Swinnen, "COVID-19 and Resilience Innovations in Food Supply Chains," in *COVID-19 & Global Food Security*, ed. J. Swinnen and J. McDermott (Washington, DC: IFPRI, 2020).

FIGURE 1 Traditional and modern integrated supply chains

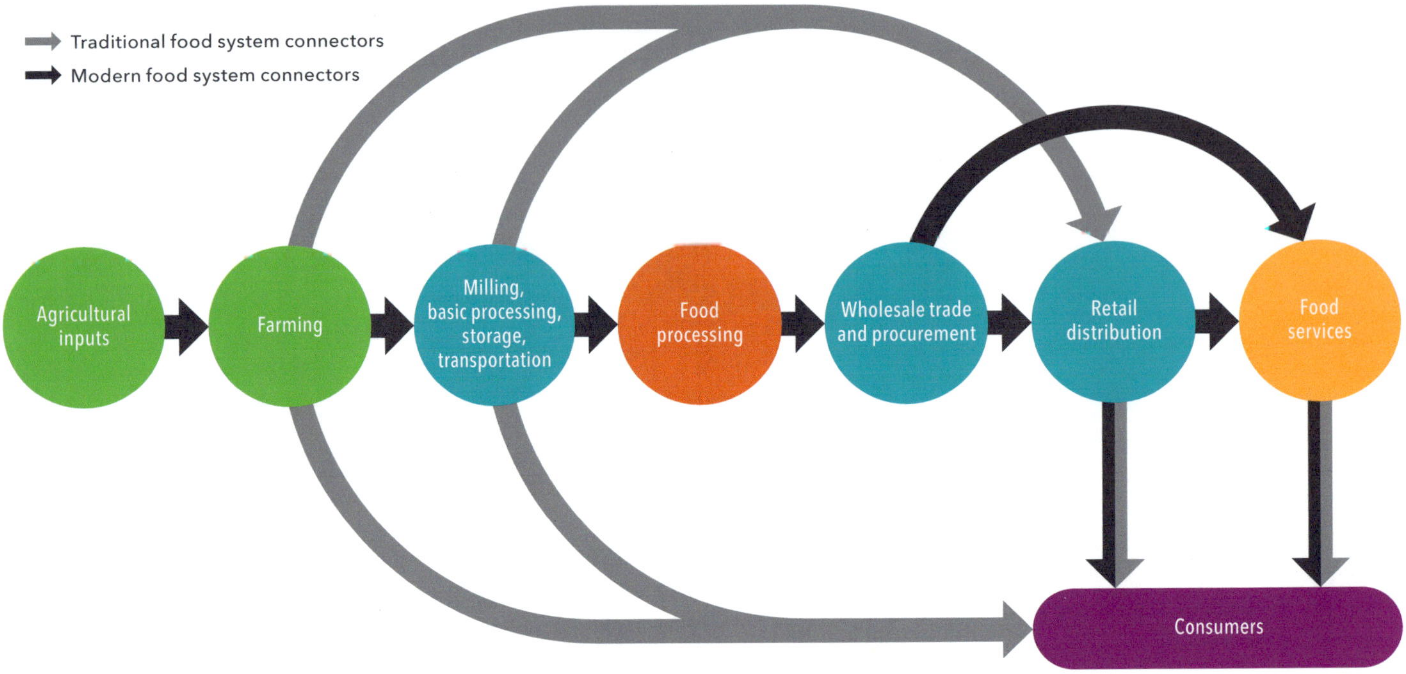

Modern supply chains (integrating all segments in Figure 1) have generally been the least affected, possessing greater capacity to adjust and innovate to keep supply chains running. Large-scale operators in modern supply chains benefited from a fair degree of control over input supplies and marketing channels, greater flexibility to switch between suppliers within their networks and between destination markets, and sufficient resources to innovate and "pivot" business operations. ("Pivoting" refers to fundamental shifts by businesses in strategy and practices in response to adverse shocks or to take advantage of major new opportunities.[3] Below, we introduce the term "co-pivoting" to refer to significant complementary shifts in business strategies and operations by firms upstream or downstream from pivoting businesses.[4])

Pivoting by private food businesses and intermediaries in 2020 typically leveraged digital platforms and/or new types of logistics business models that were already beginning to emerge before the pandemic. Such innovations in business operations, especially the use of e-commerce, e-logistics, e-payment, and e-procurement platforms to link to suppliers and buyers, proved effective in adjusting to the major supply and demand shocks provoked by the pandemic, and they are likely to endure. While adoption of these innovations may be challenging in some contexts, they provide important new opportunities, particularly for the many small and medium enterprises (SMEs) in developing-country food supply chains. Importantly, these innovations have been wholly market driven and introduced by private operators, though facilitated by existing, primarily publicly provided infrastructure as well as regulations for mobile communication technology and other connectivity.

PANDEMIC ADAPTATION STRATEGIES OF GLOBAL AND LOCAL FOOD BUSINESSES

Lockdown policies to contain the spread of COVID-19 provoked massive shocks to entire food supply chains, interrupting supplies and altering demand. The global recession has hurt nonfood activity the most, but business disruptions and severe employment and income losses have also spilled over to the food sector by interrupting input deliveries and dampening food demand. On the supply side, as downstream food businesses faced obstacles in sourcing agrifood products from farmers, they sought to ensure supplies either by strengthening their own vertically integrated delivery channels, in the case of large operations, or by using a range of new logistics and delivery intermediaries organized through e-platforms. On the demand side, partial or complete closure of restaurants and other food service businesses as well as restrictions on mobility meant that consumers began buying relatively more from retail distributors, especially modern stores that offer a broad variety of food products and other necessities. Social distancing measures and people's fears of infection also increased demand for home delivery, especially through e-commerce platforms.

ACCELERATING CHANGE

COVID-19 has accelerated many ongoing shifts in food supply chain operations. During the past 25 years, the agrifood sector underwent major and rapid changes, especially in developing countries. Financed by large-scale foreign or domestic investors, supply chains modernized, a process that includes restructuring of wholesale and retail distribution, logistics, processing, and input supply to provision rapidly expanding urban food markets.[5] Upstream innovations included modern farm inputs and new technologies; downstream innovations, our focus here, included expansion of supermarkets, franchised fast-food service, and packaged processed food. In developing countries, the characteristics of traditional markets – including fragmented supply chains, missing service and input markets, inadequate skills, and underfinanced suppliers – have tended to hamper adoption of these innovations.

As a result, two broad modernization modalities have emerged. Some large-scale operators have set up vertically integrated supply chains to control input and output delivery and limit transaction costs. Other large-scale supermarket chains, traders, and food processors employ an array of SMEs to support transportation, logistics, distribution, and delivery. Vertically integrated supply chains have shown greater resilience and capacity to adjust and innovate during the pandemic. Supply chains dominated by SMEs, common in many developing countries, have been more vulnerable. As mentioned, these systems have less capacity to adjust to restrictions on labor movements (when relying on hired labor, rather than family members) and greater susceptibility to disruptions in other input provisioning and transportation (especially where storage and processing capacities are inadequate).

SUPPLY: *Resilience and adaptation*

In some contexts, these modernization processes have led to strongly dualistic market structures, with modern vertically integrated supply chains serving one market segment and traditional SMEs serving another. Senegal's fresh fruit and vegetable supply chains illustrate the stark contrast in ability to adjust to the pandemic shock (Box 1). The vertically integrated, large-scale modern firms, which cater exclusively to export markets, have suffered little impact from the pandemic. These firms were able to adjust market channels and adapt business operations to circumvent labor restrictions. In contrast, small-scale farms, traders, and handlers operating in Senegal's poorly integrated domestic markets were severely affected by labor restrictions and disruptions in input supply, aggravated by a lack of adequate storage and limited capacity to manage risks.[6]

In Ethiopia, vegetable supply chains were also severely affected by disruptions in transport and in the supply of key farm inputs. In response to the pandemic, the government introduced trade restrictions to protect domestic producers from import competition. The impact on Ethiopian vegetable farmers was mixed. Those farmers who could sell into urban markets benefited from reduced local and international competition and higher prices, but those who could not trade to other parts of the country lost out.[7] However, Ethiopia's smaller vegetable farms were less affected by pandemic-related disruptions than medium-sized farms, as smaller farms rely less on hired labor.[8] This finding is consistent with the more general hypothesis that vulnerability to reduced labor availability, as resulted from pandemic restrictions, shows an inverted U-shaped relationship with farm size.[9] That is, small farms that rely on family labor have been largely unaffected by labor restrictions, but vulnerability increases among medium-sized farms with relatively high dependence on hired labor. Resilience has been much greater among agribusinesses large enough to benefit from significant economies of scale and financial capacity; these businesses managed to assure their labor supply by reorganizing labor shifts and arranging for safe transportation for workers, as was observed in Senegal's large export firms.

BOX 1 COPING WITH COVID-19 SHOCKS IN FRUIT AND VEGETABLE SUPPLY CHAINS IN SENEGAL

Senegal's large-scale fresh fruit and vegetable companies showed great resilience and were hardly affected by the pandemic, as documented in a recent study. These large companies operate vertically integrated supply chains oriented entirely to export markets. They were able to avoid labor supply disruptions due to COVID-19 infections by providing their workers with protective gear, doubling the number of shifts (and thus halving the number of laborers per shift), and investing in safer transport facilities to bring workers to and from the fields and collection centers. Vertical integration meant these companies also could rely on secure input supplies and marketing channels, as well as on their own cooled storage and transport capacity.

Small-scale, traditional fruit and vegetable enterprises catering to Senegal's domestic market, in contrast, were hit hard in the first months after social-distancing measures were imposed to contain the virus. These companies had little capacity to adjust and innovate in response to the shock and, consequently, were deeply affected by the mobility restrictions, closure of shops and wet markets, and their lack of access to credit and cold storage facilities. Senegal's traditional horticulture producers and traders generally faced lower output prices (caused in part by lower demand and quality loss) and rising input costs, which squeezed their profit margins and reflect the lack of resilience in local supply chains.

Source: Based on K. Van Hoyweghen, A. Fabry, H. Feyaerts, I. Wade, and M. Maertens, "The Resilience of Horticultural Supply Chains to the Covid-19 Pandemic: Insights from Senegal," *Agricultural Economics* (forthcoming, 2021).

DEMAND: *Modern retail and e-commerce*

Previous food and health safety crises led to increased supermarket purchases and declines in shopping at traditional wet markets; the SARS epidemic, for example, jumpstarted e-commerce in China. The COVID-19 pandemic likewise has increased modern grocery store sales at the expense of traditional stores (Figure 2 and Table A1). Albeit starting from low levels, e-commerce in food retail jumped by over 100 percent during 2020 in many middle-income countries, including Brazil, Indonesia, and South Africa, and by almost 50 percent worldwide. Although Figure 2 refers to consumer e-purchases at the retail stage only, use of e-commerce platforms in other segments of the supply chain such as logistics is growing even faster and fundamentally changing the structure of food business operations, as we detail further below.

BUSINESS INNOVATION FOR RESILIENCE: *E-commerce*

During the pandemic, many modern food businesses have been able to make innovations in their operations to cope with pandemic-related supply and demand shocks. Adjustment capacity and strategies have varied not only by type of supply chain but also within already modernized supply chains. Often, differences have their origin in pre-pandemic innovation strategies. Some modern food-industry firms pivoted by expanding e-commerce to reach consumers and, for inputs, by expanding e-procurement to reach processors and farmers. E-commerce had been growing pre-pandemic but got a big boost from COVID-19. Firms that had already introduced e-commerce and delivery services could ramp up these operations quickly. Those that had not established this digital and logistics capacity either pivoted to e-commerce by involving specialized intermediaries or were left behind. At the same time, new specialized "delivery intermediaries," such as third-party logistics service (3PLS) firms, have emerged and thrived using e-platforms, as illustrated in stylized fashion in Figure 3. These intermediaries engage with wholesale traders, processors, restaurants, and other food service providers, providing capacity to deliver and procure food products. Many of these companies already existed, but the pandemic led them to fast-track innovations to intensify and expand their operations. Intermediaries in wholesale, finance, and logistics rapidly moved to facilitate new business strategies building on e-procurement and e-commerce, as well as new logistics solutions that often required substantial investments.

Three current trends involving fundamental changes in individual business and supply-chain operations through digital technologies, described below, can be expected to continue to expand and evolve even after the pandemic ends.[10]

E-COMMERCE ENTERING FOOD VALUE CHAINS. During the 2010s, the food sector witnessed rapid growth of e-commerce firms that acquired, founded, or partnered with "brick-and-mortar" retailers. Notable examples include e-commerce firm Amazon buying the supermarket chain Whole Foods in the United States in 2017 and establishing Amazon Fresh stores in 2020, and China's JD.com (also a major e-commerce company) buying Yonghui Superstores in 2015. Conversely, Walmart-India acquired Flipkart, a large e-commerce firm, in 2018; and Reliance, another leading supermarket chain in India, founded Jiomart as a grocery e-commerce subsidiary in 2019. With the pandemic, these combinations of e-commerce and physical stores took off. Businesses unable to follow this strategy faltered, even modern large-scale operations, as exemplified by the case of Future Retail in India (Box 2).[11]

FOOD RETAILERS LEVERAGING E-COMMERCE TO INTEGRATE THEIR SUPPLY CHAINS. Downstream food retailers are either expanding their own e-based logistics or leveraging new intermediaries. Before the pandemic, supermarket chains in developing countries had begun to operate online delivery services

FIGURE 2 Growth rates of retail food purchases by type of provider in middle-income countries, 2019–2020

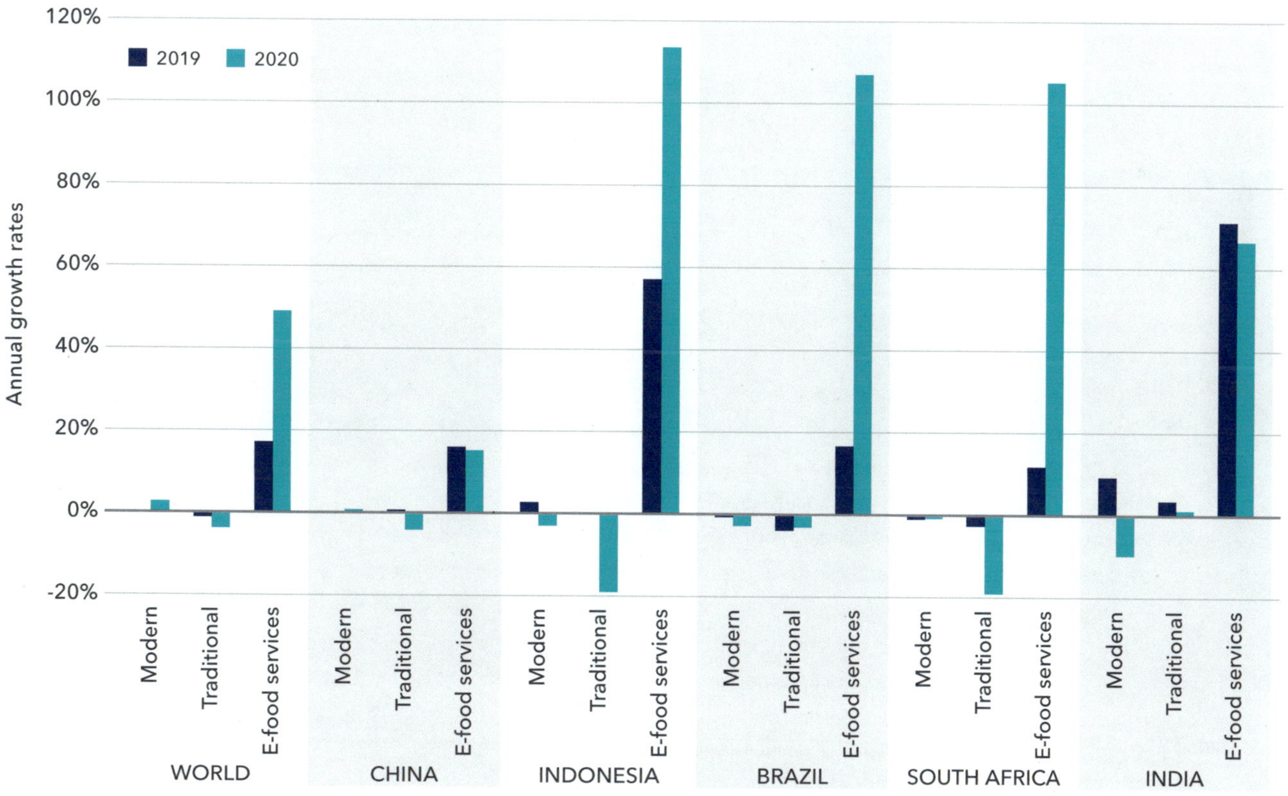

Source: Based on data from Euromonitor International, accessed January 20, 2021; for details see Table A.1.

Note: Growth rate is for real per capita food purchases at retail level. "World" data cover 103 countries. "Modern" retail stores include convenience stores, supermarkets, hypermarkets, and discounters. "Traditional" grocery retailers are those that are "non-chained," small-scale stores owned by families, and/or run on an individual basis, and do not include informal retailers in open markets or street vendors.

FIGURE 3 Pivoting with e-intermediaries in response to COVID-19 shocks

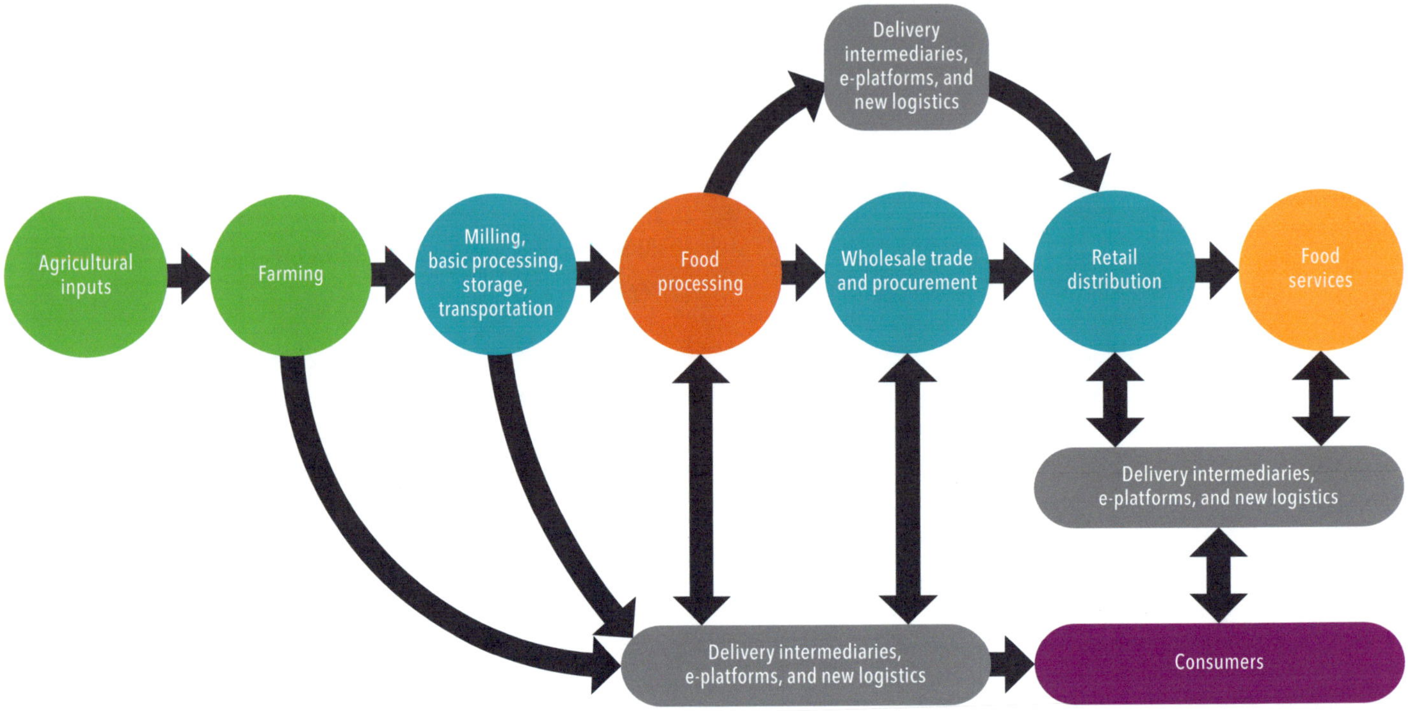

BOX 2 FALL AND RISE OF MODERN FOOD RETAILERS IN INDIA

Many brick-and-mortar retailers that did not shift to e-commerce before or early on in the COVID-19 crisis lost substantial business. Future Retail, for instance, was a leading retailer (including of food) in India, with 1,800 stores in 420 cities and wholesale and logistics divisions with trucks and distribution centers around the country. In the five years preceding the pandemic, Future Retail rapidly expanded sales locations and product assortment to compete with Reliance, Walmart-Flipkart, and other supermarket chains. Doing so left the company heavily indebted. Future Retail did not move into e-commerce, and profits plummeted when lockdown regulations and consumer fear suddenly curtailed in-store shopping. By August 2020, the company was on the verge of bankruptcy and subsequently was taken over by Reliance. By acquiring Future Retail's large network of stores and supply-chain logistics, Reliance greatly increased its market share, fortifying its combination of brick-and-mortar retail, online commerce, and logistics capacity.

The Future Retail case is an eye-opening reminder to businesses to continually update their strategies. Future Group was an early mover in supermarkets in India, and had been highly innovative in regional chain acquisitions, format diversification (such as very cheap mini-stores to penetrate poor neighborhoods), manufacturing of private-label food products, and development of logistics operations. But it failed to make the one major shift needed to make it resilient to the COVID-19 shock. As a result, the pandemic led to further consolidation of India's food retail sector, leaving just three clear market leaders, Reliance, Walmart-Flipkart, and Amazon.

Source: Based on T. Reardon, A. Heiman, L. Lu, C.S.R. Nuthalapati, R. Vos, and D. Zilberman, "'Pivoting' by Food Industry Firms to Cope with COVID-19 in Developing Regions: E-commerce and 'Co-pivoting' Intermediaries," *Agricultural Economics* (forthcoming, 2021).

alongside in-store retail as a means to expand market share and outcompete small family-owned stores and traditional markets. In response to the lockdowns and social distancing, leading retail and fast-food chains greatly expanded home delivery and curbside pickup. Some retailers did this by enhancing their own logistics platforms and those of their subsidiaries. For example, in India, Walmart-Flipkart drew on the services of its own Ekart Logistics, both to make its own deliveries and to sell logistics services to other e-commerce and brick-and-mortar retail firms.[12]

Where supermarkets and e-commerce firms relied on intermediaries for logistics services, the logistics companies have responded either by leveraging their existing e-commerce capacity or "co-pivoting" to meet the changing delivery needs of their clients. In some cases, these firms obtained investment financing from retailers to facilitate this change in business orientation. For example, India's Walmart-Flipkart invested in the logistics startup Shadowfax, an e-platform that links local logistics SMEs to e-commerce companies.[13] When India's medium and large supermarket chains suddenly needed to meet growing online demand and expand delivery capacity in 2020, Flipkart could draw on Shadowfax, while also bolstering its own logistics services. Another example of co-pivoting in India is the response of e-commerce firm BigBasket to labor constraints on delivery operations after urban drivers returned to their villages during the country's lockdown. To address this, BigBasket partnered with Uber-India and Rapido (an online bike-taxi firm) in April 2020. Uber, in turn, added food delivery to its online services, and Uber and Rapido expanded their fleets to deliver BigBasket's increased online food orders to customers.[14]

PROLIFERATION OF NEW DELIVERY INTERMEDIARIES FACILITATING E-COMMERCE IN FOOD SUPPLY CHAINS. Uber and Rapido are part of the broader trend of using intermediaries, which accelerated with the pandemic. E-intermediaries provide an app that allows consumers to select products from a list of subscribed retailers. The delivery intermediary then either fetches and delivers the product or enlists a third-party provider to do so. Rappi in Latin America and Swiggy in India are examples of large-scale food delivery services that expanded rapidly during the 2010s by adding e-commerce to their operations, initially to increase market shares, and expanded further in 2020 as a crucial adjustment to changing markets during the pandemic.[15]

In much of lower-income Asia, Africa, and Latin America, SMEs still dominate food retail and services. During the pandemic, as small shops and restaurants were severely affected by mobility restrictions and consumer fears, these businesses pivoted to deliver food products and meals using online platforms. Financial and marketing intermediary services co-pivoted to facilitate many of these changes. In Thailand, for instance, SME retailers began selling food directly to consumers via Facebook, cellphone networks, and local SME delivery apps. Also, large retailers such as Reliance in India are facilitating "local" e-commerce by SME retailers through Jiopay and Jiomart (Reliance's e-payment and e-commerce divisions). These changes started before the pandemic but rapidly accelerated in 2020, providing new opportunities for small and medium food businesses and delivery intermediaries.

WHAT POLICIES CAN DO AND WHAT THEY SHOULD NOT DO

The coronavirus has posed a major global health threat and caused a massive global economic shock. Food supply chains were not exempt from the consequences of the virus or of measures taken to contain its spread. Yet, those supply chains have shown remarkable resilience. The agrifood sector was widely declared an "essential" sector during the pandemic, which facilitated the continued flow of food supplies. Nevertheless, the sector did experience major disruptions caused by demand shocks, labor restrictions, market closures, and, in some instances, trade restrictions. Evidence from China shows that such disruptions could be limited by creating "green lanes" that exempt transport, production processes, and distribution of agricultural inputs and food products, as well as movements of food-sector workers, from lockdown measures.[16] However, such measures would naturally be less effective where supply chains are poorly integrated and where SME food businesses also faced large demand shocks.

Agrifood businesses, large and small, in modern, integrated supply chains have shown greater resilience because of their greater capacity and opportunities to adjust or pivot business operations, which in turn has accelerated ongoing food system change. Their adjustments included introduction of flexibility in labor access, in product procurement, in marketing, in technology (especially the use of e-platforms), and in financial resilience. We expect continued diffusion of these organizational innovations across food supply chains, including in developing countries. Many of the innovations, such as e-platforms for logistics, marketing, credit, and payments, were initiated by large firms. In some cases, innovations deepened the vertical integration of single food businesses at the expense of smaller players and of competition. Elsewhere, and significantly, pivoting by large companies provided new opportunities for SME intermediaries in logistics, retail, and wholesale trade, and also for SMEs in processing and farming itself. The expansion of e-commerce has helped SMEs deliver food to consumers under lockdowns and other constraints and enhanced the resilience of supply chains in developing countries. Its expansion can be expected to continue after the pandemic.

These food system innovations have foremost been market responses by private food businesses adjusting to the changing demand and purchasing behavior induced by social distancing measures, though their opportunities to adjust depended heavily on the basic infrastructure, mobile ICT networks, and regulations put in place by past public investment and policies. This type of public support is essential for market integration and lowering transaction costs along supply chains, allowing for food business operations to innovate and pivot. Governments would be ill-advised to organize supply chain integration directly, but rather should focus on their role as facilitators and provide targeted support to improve access to affordable mobile services and ease market entry for SMEs to ensure that business pivoting and changing practices do not lead to increased concentration of food markets. With this support, food supply chain modernization and innovation can contribute to decent livelihoods and employment generation along supply chains while enhancing resilience to the impacts of future shocks and disruptions, and thus contributing to the transformation of food systems.

APPENDIX

TABLE A1 Real per capita food purchases at retail level by type of provider, 2015–2020

	PER CAPITA RETAIL FOOD SALES (US$ at constant prices)						ANNUAL GROWTH RATES				
	2015	2016	2017	2018	2019	2020	2016	2017	2018	2019	2020
WORLD											
Modern	518.5	520.5	521.5	519.4	518.8	532.8	0.4%	0.2%	-0.4%	-0.1%	2.7%
Traditional	389.1	387.1	384.3	385.0	380.3	366.0	-0.5%	-0.7%	0.2%	-1.2%	-3.8%
E-food services	11.3	13.6	16.1	18.6	21.8	32.5	20.4%	18.4%	15.5%	17.2%	49.1%
CHINA											
Modern	335.6	333.9	337.1	333.3	333.2	336.0	-0.5%	1.0%	-1.1%	0.0%	0.8%
Traditional	672.0	672.2	656.1	629.9	634.2	608.0	0.0%	-2.4%	-4.0%	0.7%	-4.1%
E-food services	19.9	25.7	30.4	36.0	41.8	48.2	29.1%	18.3%	18.4%	16.1%	15.3%
INDONESIA											
Modern	66.1	69.1	70.2	72.2	74.2	72.0	4.5%	1.6%	2.8%	2.8%	-3.0%
Traditional	330.5	336.7	342.8	347.1	348.2	282.2	1.9%	1.8%	1.3%	0.3%	-19.0%
E-food services	0.2	0.4	0.9	1.4	2.2	4.7	100.0%	125.0%	55.6%	57.1%	113.6%
BRAZIL											
Modern	152.9	150.8	147.9	142.8	141.8	137.8	-1.4%	-1.9%	-3.4%	-0.7%	-2.8%
Traditional	127.5	109.2	104.6	99.6	95.7	92.7	-14.4%	-4.2%	-4.8%	-3.9%	-3.1%
E-food services	1.1	1.1	1.1	1.2	1.4	2.9	0.0%	0.0%	9.1%	16.7%	107.1%
SOUTH AFRICA											
Modern	370.5	368.8	367.5	366.0	362.1	358.7	-0.5%	-0.4%	-0.4%	-1.1%	-0.9%
Traditional	230.4	232.6	239.1	238.6	232.1	187.7	1.0%	2.8%	-0.2%	-2.7%	-19.1%
E-food services	1.3	1.4	1.5	1.7	1.9	3.9	7.7%	7.1%	13.3%	11.8%	105.3%
INDIA											
Modern	5.7	6.3	6.9	7.5	8.2	7.4	10.5%	9.5%	8.7%	9.3%	-9.8%
Traditional	248.9	257.5	270.8	280.5	290.7	294.4	3.5%	5.2%	3.6%	3.6%	1.3%
E-food services	0.2	0.4	0.5	0.7	1.2	2.0	100.0%	25.0%	40.0%	71.4%	66.7%

Source: Based on data from Euromonitor, accessed January 20, 2021.

Note: "World" data cover 103 countries. "Modern" retail stores include convenience stores, supermarkets, hypermarkets, and discounters. "Traditional" grocery retailers are those that are "non-chained," small-scale stores owned by families and/or run on an individual basis, and do not include informal retailers in open markets or street vendors.

REGIONAL DEVELOPMENTS

AS THE CORONAVIRUS PANDEMIC REACHED EVERY CORNER OF THE WORLD IN 2020, countries responded rapidly with an array of policies to stop the spread of the highly contagious disease, and then with social and economic policies to protect food security, incomes, and livelihoods. This experience brought attention to weaknesses in health, economic, and social protection systems. But it also showcased new innovations, policy approaches, and the surprising resilience of food systems. The diverse experiences of the world's major regions have important lessons for creating sustainable, equitable, efficient, healthy, and resilient food systems. The impact of COVID-19 on food systems, wellbeing, and future transformation is examined for each major region:

- Unprecedented scale-up of social protection programs in Africa south of the Sahara
- Falling household incomes for rich and poor in the Middle East and North Africa
- Central Asia's risky reliance on remittances
- Impacts of South Asia's stringent lockdowns on migrant and informal labor
- East and Southeast Asia's interregional trade expansion and economic recovery
- Urbanization, obesity, and COVID-19 vulnerability in Latin America and the Caribbean

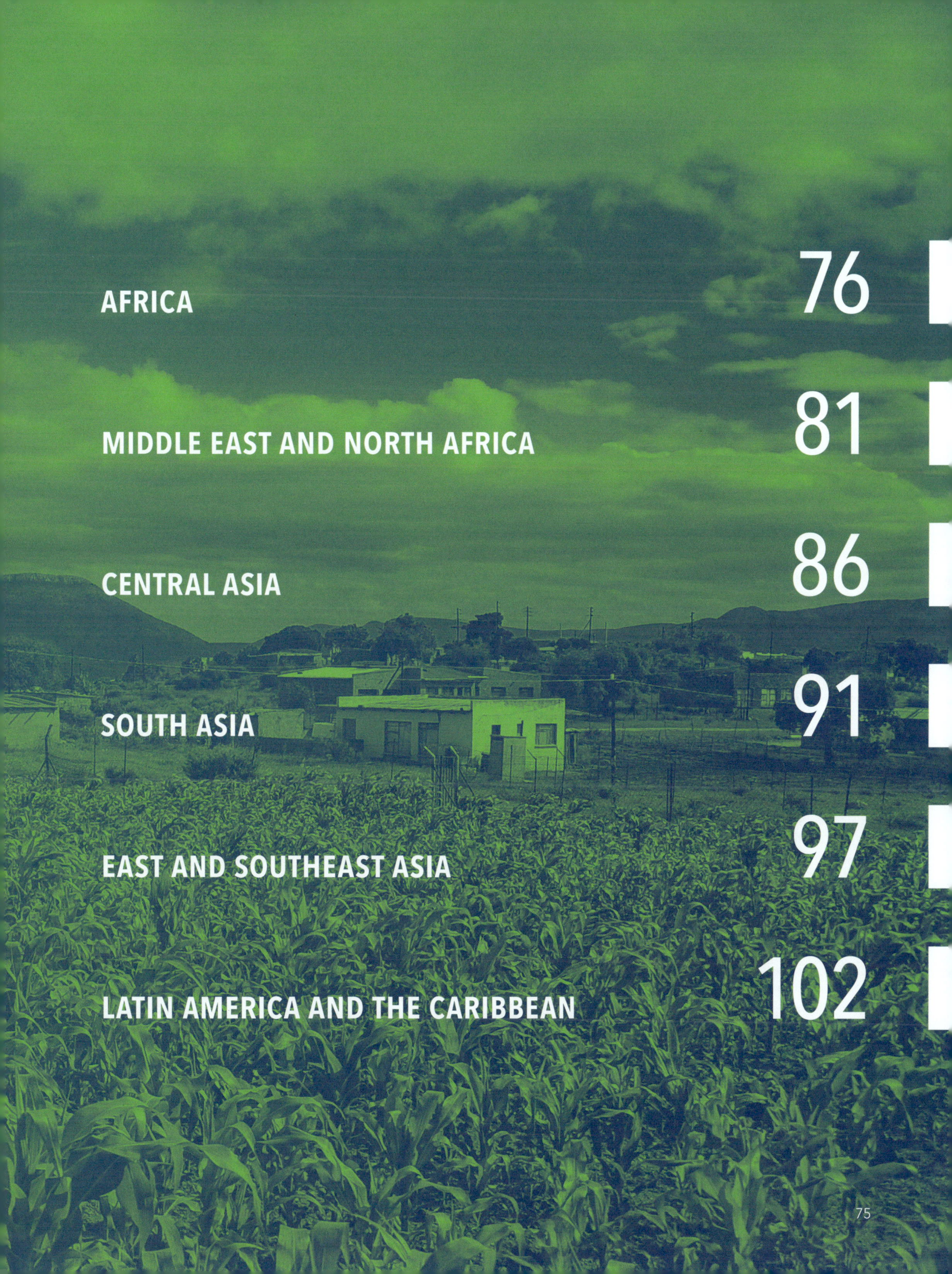

AFRICA

SAMUEL BENIN, KWAW ANDAM, AND JOHN ULIMWENGU

Samuel Benin is deputy division director for the Africa Region, International Food Policy Research Institute (IFPRI), Washington, DC. **Kwaw Andam** is a research fellow and country program leader of the Nigeria Strategy Support Program, IFPRI, Abuja. **John Ulimwengu** is a senior research fellow, Africa Region, IFPRI, Washington, DC.

Africa south of the Sahara has so far escaped the direst health impacts of the coronavirus pandemic, but short-term policy responses and the global slowdown have had major impacts on growth, value chains, incomes, trade, poverty, and consumption. The medium- to long-term impacts on nutrition and health will depend on the rate of recovery. To build greater resilience to future shocks, African countries will need to take a food systems approach to agrifood policy planning and implementation that includes better anticipation of risks and preparedness for shocks. They must also find ways to expand limited fiscal resources to make investments that embrace inclusion and promote good governance and accountability in all parts of the system.

SHORT-TERM IMPACTS OF COVID-19

The health effects of the COVID-19 pandemic in Africa south of the Sahara have been less severe than first anticipated given the region's overall fragility. With 14 percent of the world's population, Africa had reported less than 5 percent of global confirmed cases of COVID-19, and about 89,000 pandemic deaths as of the end of January 2021.[1] Although there may be some underreporting of cases and deaths, there is growing evidence that the low death rate is largely explained by Africa's sizable young population, along with other factors including climate, genetics, and behavioral differences.[2]

However, as in other regions, governments across Africa implemented a variety of lockdown policies in early 2020 to combat the spread of the virus. These policies varied in terms of their geographic coverage (many were restricted to outbreak-affected urban areas, but some applied nationwide), the types of socioeconomic activities that were restricted (based on essential versus nonessential classifications), the start of the lockdowns (mostly in March, with Rwanda being the first, but some, including Botswana, Eritrea, and Malawi, not until April), and their duration (ranging from two to three weeks in Malawi to nearly three months in South Africa). These measures had substantial impacts on economic growth, poverty, and other key socioeconomic indicators. In general, lockdown periods coincided with the agricultural seasons in central, eastern, and western Africa, but began at the end of the season in southern Africa.

Protests over lockdowns broke out in a number of countries and hint at how the public perceived their respective governments' motivations for the lockdown policies – for example, to prolong power or quash opposition (Malawi and Ethiopia) or to target certain social groups (Uganda). Public outcry over lack of evidence to support the policies, as well as violent attacks on citizens by security forces enforcing the lockdown policies, which led to deaths in countries including Kenya and South Africa, contributed to modification or reversal of the policies in a number of countries.[3]

Several ex ante analyses of the short-term impacts of Africa's lockdown policies, conducted and updated throughout 2020, projected an economic recession, interrupting nearly 25 years of sustained economic growth in the region.[4] Estimates of the contraction in 2020 GDP for Africa as a whole range from 1.7 percent under a baseline scenario (mild spread of the disease in the first half of 2020, with lockdowns ending by the middle of the year) to as much as 5.1 percent under a catastrophic scenario (severe spread of the disease with a high number of cases and lockdowns extending beyond the middle of 2020).[5] The drop in GDP stems in part from the contraction of global demand for the primary commodities produced by African countries, as global production, travel, and trade were disrupted.[6]

Household incomes also fell as a result of reduced employment during lockdowns and reduced remittances from outside the region. Country-level economy-wide models show declines in household income in a number of countries, and GDP losses ranging from 9.7 percent in Mali to 38.0 percent in Rwanda (Figure 1).[7] Phone surveys of households and firms in the food supply chain conducted in late 2020 in Burkina Faso, Ethiopia, and Nigeria confirm these modeled impacts and pathways, for example, showing employment and income losses as firms closed their operations.[8]

In terms of food systems, disruptions in value chains and informal urban trade caused shortages and higher food prices, as observed, for example, in Rwanda and the Democratic Republic of the Congo.[9] Overall, food shortages were caused by closure of wet markets, restrictions on street vendors, and buying frenzies and hoarding ahead of the lockdowns. Studies from Ethiopia show there have been differences in the impacts across value chains. Although Ethiopia's food value chains in the aggregate may have been resilient to the pandemic shock,[10] the dairy sector (especially raw milk vendors and small dairy shops) was hit hard as demand for its products slumped and prices of feed increased by 30 to 40 percent and those of veterinary services by 15 to 20 percent.[11]

Lockdown measures, lost income, and perceptions of disease risk have increased poverty and altered diets. Increases in poverty rates during the lockdown range from 3.8 percent in South Sudan to 15.0 percent in Senegal (Figure 1), and for the region as a whole, extreme poverty is expected to increase by up to 2.8 percent, representing an additional 37.5 million people.[12] Phone surveys of households in Ethiopia show that consumption of raw or uncooked vegetables and dairy products declined due to fear of coronavirus transmission, but overall calorie consumption increased compared to pre-pandemic periods.[13] Not surprisingly, the increase in calorie consumption was significantly greater among households that did not experience a loss in income than among those that did.

POLICIES AND OTHER RESPONSES

Beyond the lockdown strategies, governments across Africa also adopted macro-level policies – including fiscal, monetary, exchange rate, and balance of payment policies – to mitigate the socioeconomic impacts of the pandemic and the health-related restrictions. For example, the West African Economic and Monetary Union (WAEMU) temporarily suspended its rule restricting countries' fiscal deficits to 3 percent of GDP in order to allow member states to increase their budget deficit ceiling.[14] In addition to these macro-level instruments, there were supports targeted to specific sectors (for example, health, tourism and hotels, and agriculture), businesses (especially the informal sector), and households (especially the vulnerable and recently laid-off), including direct transfers to businesses and households. Other pandemic relief policies included reducing or fixing prices of inputs and grain as well as reducing, postponing, or canceling various tax payments, fees, and utility bills. Kenya offered complete tax relief to those earning less than 24,000 Kenyan shillings (US$220) per month, Malawi allowed a six-month postponement of tax payments, and Rwanda fixed prices for essential pandemic-related commodities such as hand sanitizer, gloves, face masks, and fuel.[15]

Most notably, social protection programs were scaled up substantially. Estimates suggest cash transfer programs, for example, were reaching 11 percent of the population in African countries in July 2020, up from just 3 percent before the crisis.[16] Despite this increase, the average coverage rate in Africa remains by far the lowest among developing regions, as the average coverage rate of cash transfer programs in response to COVID-19 has reached 38 percent across all developing regions. But Africa's rate of increase is the second highest, with countries including the Republic of the Congo, Guinea, Kenya, Mali, Togo, and Zimbabwe surpassing the average rate of increase for all developing countries.[17]

The cost of the various policies – in terms of higher government expenditure or foregone government revenue – is estimated at about $38.5 billion or 2.4 percent of GDP for African countries.[18] However, the flexibility of governments in their spending choices (termed fiscal space) appears limited, as the estimated liquidity support (equity injections, loans, asset purchase or debt assumptions) is less than 50 percent of the total cost, indicating potential for a future fiscal crisis in some countries, including Ethiopia, Ghana, Mali, and Senegal (Figure 2).[19] The limited availability

FIGURE 1 Impacts of pandemic lockdowns on poverty and GDP in selected African countries, 2020

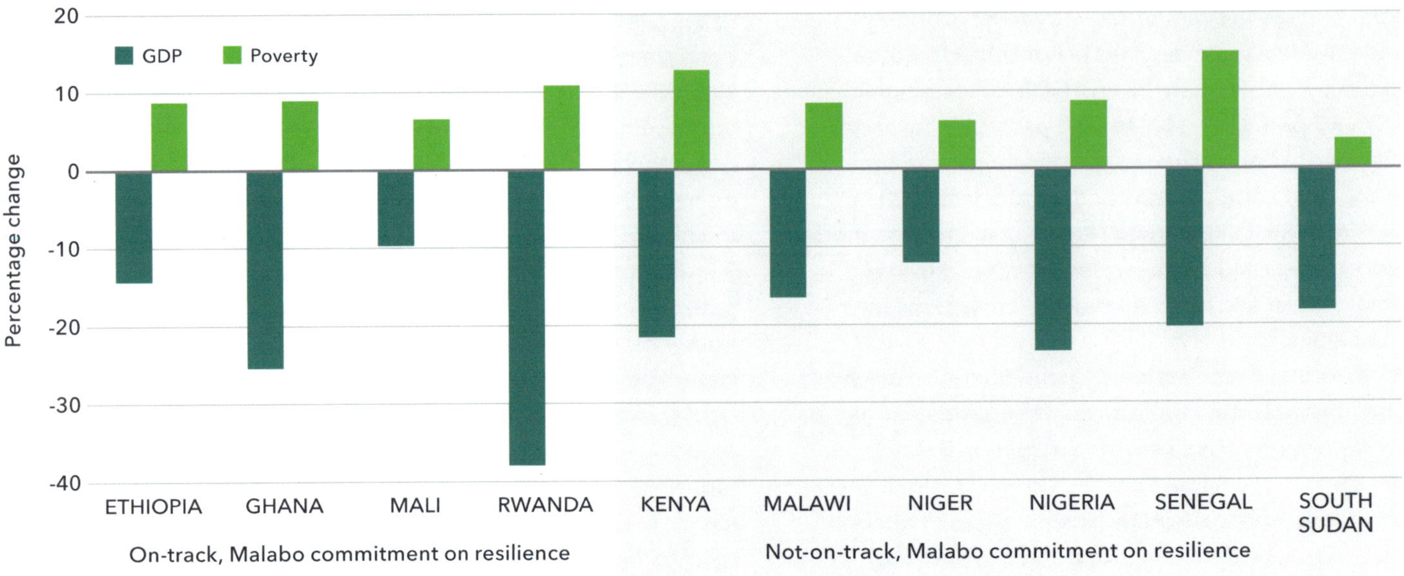

Source: Based on S. Amewu et al., "The Economic Costs of COVID-19 in Sub-Saharan Africa: Insights from a Simulation Exercise for Ghana," *European Journal of Development Research* 32 (2020): 1353–1378; K.S. Andam et al., "Impacts of COVID-19 on Food Systems and Poverty in Nigeria," in *Advances in Food Security and Sustainability*, vol. 5, ed. Marc J. Cohen (Cambridge, MA: Elsevier, 2020); C. Arndt et al., "Covid-19 Lockdowns, Income Distribution, and Food Security: An Analysis for South Africa," *Global Food Security* 26 (2020): 100410; African Union Commission, *Second Biennial Review Report of the African Union Commission on the Implementation of the Malabo Declaration* (Addis Ababa: 2020).

Note: "Poverty" is the headcount poverty ratio. Changes in GDP (gross domestic product) and poverty are for the lockdown periods, which vary by country.

FIGURE 2 Fiscal measures in response to COVID-19 in selected African countries, 2020

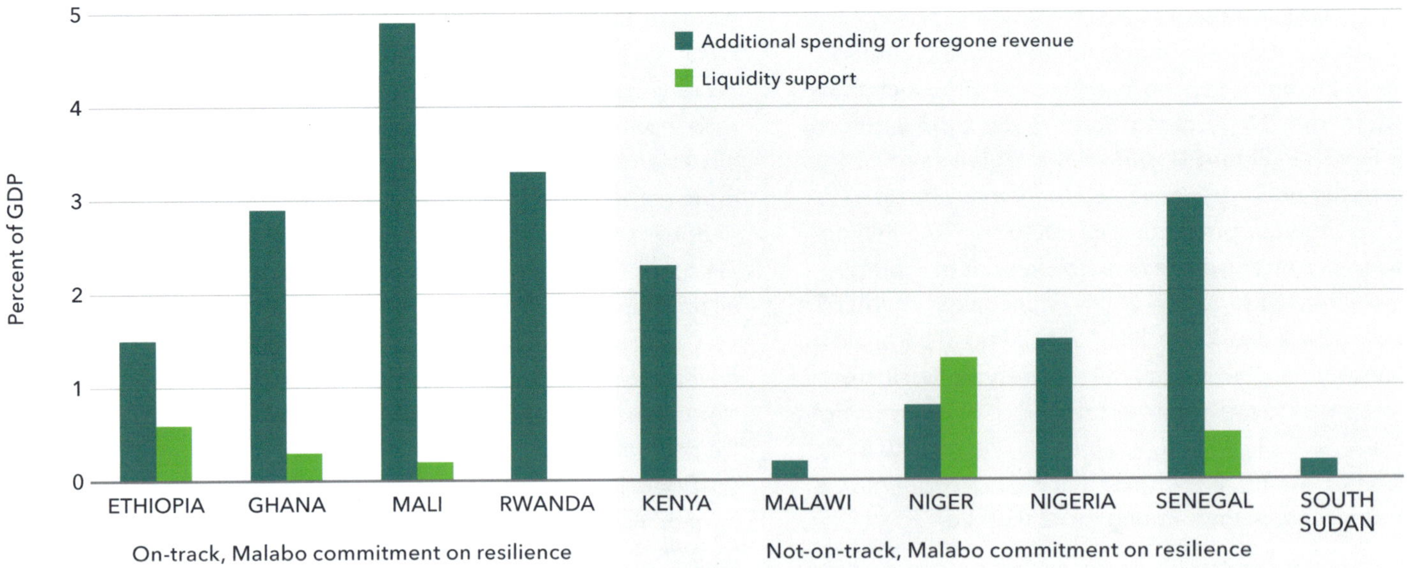

Source: Based on African Union Commission, *Second Biennial Review Report of the African Union Commission on the Implementation of the Malabo Declaration* (Addis Ababa: 2020); IMF, "Fiscal Monitor Database of Country Fiscal Measures in Response to the COVID-19 Pandemic," October 2020.

Note: "Liquidity support" includes equity injections, loans, asset purchases, debt assumptions, and contingent liabilities; data are not available for Kenya, Malawi, Nigeria, Rwanda, and South Sudan.

of government fiscal resources to respond to such shocks is reflected in the region's slow progress toward achieving the Malabo Declaration commitments to enhancing resilience of livelihoods and production systems to climate variability and related risks.[20] Only nine countries (Burundi, Cabo Verde, Ethiopia, Ghana, Mali, Mauritania, Rwanda, Seychelles, and Uganda) are currently on-track to achieve their commitments.[21] Governments of the on-track countries seem to have greater fiscal space to implement measures to deal with the pandemic, in terms of additional spending or foregone revenue (Figure 2).

Private sector responses to the pandemic have varied, depending on the industry or business sector, whether or not they are considered essential under the lockdown classifications, and the level of government support received. Invariably, a significant portion of firms suspended investments, laid off staff, reduced production, canceled orders, and terminated or renegotiated leases. Many adapted by changing their operational practices or their products and services. In the Democratic Republic of the Congo, for example, almost 10 percent of enterprises were forced to sell assets, close to 40 percent shifted some staff to telework, and 10 percent shifted to new business activities.[22] Notable innovations in response to the pandemic are digital solutions in different sectors such as health, including rapid diagnostic testing kits, mobile testing booths, and online platforms linking patients to doctors; and trade, including mobile payment platforms for informal traders and online platforms for small and medium enterprises to sell their products.[23]

TOWARD FOOD SYSTEM RESILIENCE AND TRANSFORMATION

The reach and effectiveness of these responses will determine the pace of recovery, as well as the extent to which the short-term effects persist into the medium and long term. The COVID-19 pandemic is just one of multiple shocks that have hit the continent in recent years – including the Ebola epidemic in Guinea, Liberia, and Sierra Leone in 2014; the fall armyworm invasion since 2017; and the locust infestation in eastern Africa in 2020. In addition, human population growth, growing demand for animal-sourced foods due to urbanization and rising incomes, and increasing travel connections with Asia are changing Africa in fundamental ways that also increase the risk of disease transmission from animals to people.[24]

Given the region's vulnerability, ensuring that African countries are prepared to withstand future shocks will be central to the transformation of Africa's food systems. Accordingly, building resilience at all levels will require a food-systems approach to agrifood policy planning and implementation. Such an approach must be based on evidence from Africa's own situation and dynamics with respect to long-term issues for food system transformation such as climate and biodiversity, demography, urbanization, value chains, and trade. At the national level, this must include better anticipation of risks and preparedness for shocks, integrating budget lines for disaster preparedness and emergency response into medium-term expenditure frameworks, institutionalizing fiscal rules and stabilization funds for flexibility in managing shocks, and promoting good governance and accountability in all parts of the system.[25]

As more frequent shocks, including those related to climate change, and subsequent emergency responses demand more of the public budget, increasing the efficiency of public spending will be critical. Investments should be ramped up in key areas for long-term productivity and growth – including investments in infrastructure, early warning and monitoring and evaluation systems, agricultural R&D and extension, education, and health. The investments must embrace inclusion, as shocks and policies for dealing with pandemics affect women, men, and children differently, particularly in rural areas.[26] Such investments for the long-term will help mitigate the impacts of shocks overall and facilitate more effective and efficient responses.

To increase these public investments, African governments must find sustainable ways to broaden their revenue base and increase the efficiency of revenue collection. One critical action will be supporting informal sector actors to transition into the formal sector. Enabling their access to technologies and financial facilities will help them to raise their productivity and business growth and increase employment, all of which will add to the available revenue needed for investment for the future.

AFRICA SOUTH OF THE SAHARA COVID-19 TIMELINE, MARCH 2020–JANUARY 2021

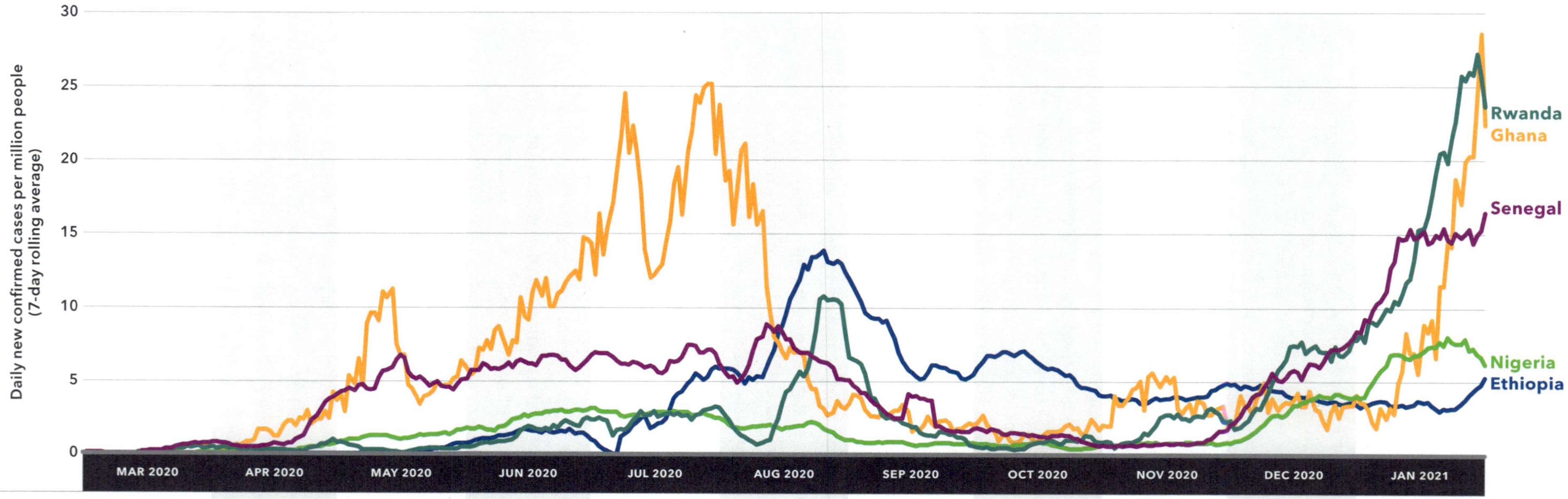

LOCKDOWNS AND HEALTH MEASURES

- **Rwanda**, Kenya, **Nigeria**, **Senegal**, South Africa and others issue lockdown orders
- Many countries close borders and ban flights
- **Ghana**, **Nigeria**, and others ban public gatherings, close schools
- South Africa's hospitals offer free testing
- **Ghana** announces coronavirus response plan
- **Nigeria** announces fund to upgrade healthcare infrastructure
- **Rwanda** uses drones to deliver coronavirus information
- **Ghana** locks down major metro areas, launches tracker app
- Cameroon begins door-to-door testing
- **Ethiopia** releases vulnerable prisoners
- Zambia introduces targeted lockdown of Nakonde
- **Rwanda** imposes targeted lockdown in Kigali
- South Africa enters vaccine trial

STIMULUS

- **Ghana** grants tax waiver for frontline workers
- **Rwanda** receives IMF loan
- **Ghana** launches support scheme for micro, small, and medium businesses
- **Rwanda** announces Economic Recovery Fund
- Zambia issues COVID-19 bond

SOCIAL PROTECTION

- **Nigeria** begins cash transfers and food aid
- Namibia provides cash transfer to those affected by lockdown
- **Ghana** approves funds for social protection
- **Rwanda** distributes food to urban poor who cannot work
- **Senegal** provides cash transfers, food aid, and utility bill support
- **Ghana** grants free water supply to all households, reduces electricity costs for poorest
- Zambia provides cash transfers to affected urban households

FOOD SECTOR INTERVENTIONS

- Kenya allows only takeout from restaurants
- South Africa exempts supermarkets from lockdowns

Source: For COVID-19 data, Johns Hopkins University, CSSE Database (updated March 8, 2021). For policy data, IFPRI, COVID-19 Policy Response Portal (accessed March 2021) and news reports.
Note: Countries and policies included here were selected to provide an indication of the types of actions taken in the region.

MIDDLE EAST AND NORTH AFRICA

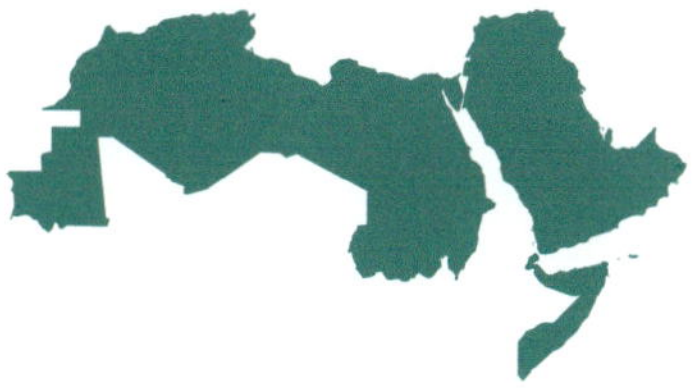

KIBROM ABAY, CLEMENS BREISINGER, DALIA ELSABBAGH, HOSAM IBRAHIM, AHMED KAMALY, AND MARIAM RAOUF

Kibrom Abay is a research fellow, **Clemens Breisinger** is a senior research fellow and leader of the MENA Regional Program and the Egypt Strategy Support Program, **Dalia Elsabbagh** and **Hosam Ibrahim** are senior research assistants, and **Mariam Raouf** is a senior research associate, all with the Development Strategy and Governance Division, International Food Policy Research Institute (IFPRI), Cairo. **Ahmed Kamaly** is deputy minister, Ministry of Planning and Economic Development of the Government of the Arab Republic of Egypt.

As in most regions, the coronavirus pandemic evolved rapidly in the Middle East and North Africa (MENA), starting as a health crisis in mid-March and later escalating into a large-scale economic crisis affecting the lives and livelihoods of millions of people. Beyond the devastating health impacts and death tolls from COVID-19, the pandemic has added to other serious challenges currently facing the region. These include low oil prices for oil-exporting countries like the Gulf countries and Libya; ongoing conflicts and political transitions in many countries; and natural disasters such as the locust swarms in Sudan and floods in Yemen. In particular, the pandemic continues to test the functioning of national food systems and expose the vulnerabilities that come with the heavy dependency of most MENA countries on food imports.

All national economies in the region have experienced severe disruptions. The impacts vary across countries and sectors, reflecting differences in both the spread of the pandemic and government responses.[1] Relative to previous quarterly output, the pandemic has caused GDP losses ranging from 1.1 percent expected in Egypt to 23.0 percent in Jordan during April–June 2020.[2] The fall in Egypt's GDP is relatively small in part because Egypt avoided full lockdown measures during the initial wave of the pandemic and quickly invested in a massive stimulus program. In Yemen, the fall in remittances resulting from the pandemic led to an estimated drop of more than 9 percent in food system outputs and household income reductions of 12.5 percent.[3]

AGRIFOOD SYSTEM RESILIENCE, BUT WITH WINNERS AND LOSERS

Agrifood systems have proven relatively resilient in the MENA region, with some value chains even benefiting from the crisis. The coronavirus outbreak coincided with some agricultural harvest seasons, including most cereal crops in Egypt and potatoes and wheat in Jordan. Reductions in agrifood GDP ranged from about 3 to 9 percent in Egypt, Yemen, and Sudan to about 38 percent in Jordan. Agriculture remains the least affected sector in the Egyptian and Sudanese economies.[4] Nevertheless, pandemic-related lockdowns and curfews in Egypt reduced the production, distribution, and sale of perishable food products such as dairy, particularly affecting smallholders who lack adequate storage and processing facilities for these products.[5] However, some value chains, including those for medicinal and aromatic plants, have enjoyed a substantial boost, fueled by increased demand for exports.

The large decline in the agrifood sector in Jordan reflects a significant fall in exports as well as labor restrictions affecting farming activity. Jordan's agriculture sector is dominated by fruit and vegetable production for export and relies heavily on foreign workers, who constitute most of the country's agricultural labor force – about 85 percent of livestock labor and 92 percent of crop labor in 2015.[6] The mobility of these workers has been hampered by COVID-19-related restrictions, and many returned to their home countries when their work licenses were not renewed.

FIGURE 1 Change in agrifood system GDP, April–June 2020 (compared with April–June 2019)

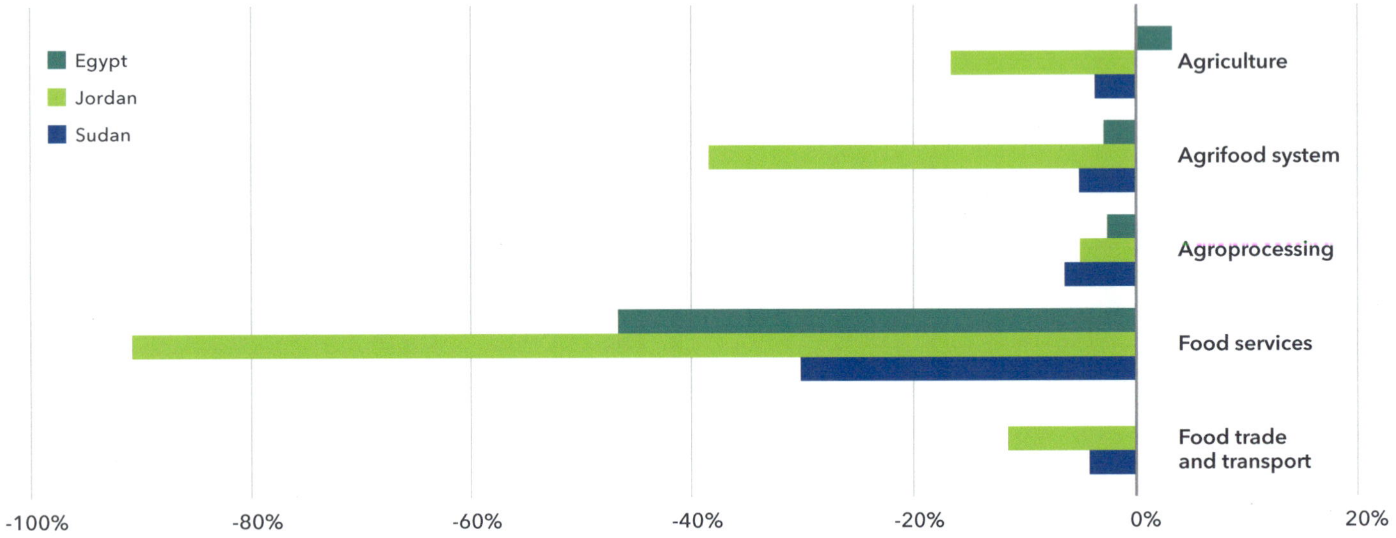

Source: IFPRI social accounting matrix (SAM) multiplier models, C. Breisinger et al., *Impact of COVID-19 on the Egyptian Economy: Economic Sectors, Jobs, and Households*, MENA Policy Note 6 (Cairo: IFPRI, 2020); M. Raouf, D. Elsabbagh, and M. Wiebelt, *Impact of COVID-19 on the Jordanian Economy: Economic Sectors, Food Systems, and Households*, MENA Policy Note 9 (Cairo: IFPRI, 2020).

The impacts of these shocks have also varied across different actors within each value chain, mainly because of their varying capacities to respond to the pandemic and associated dynamics in demand for and supply of products.[7] For example, in Egypt the pandemic has affected milk producers disproportionately more than milk processing factories.[8] Recent studies show that global demand for digital tools has significantly increased during the pandemic, including in Egypt, and those sectors that are sufficiently digitalized have been less affected by the pandemic.[9]

The greatest economic damage to the region's agrifood systems has occurred in food services, including hotels and restaurants. In contrast, losses in the food-processing sector have been relatively small and largely driven by indirect effects, including the reduced demand from restaurants and hotels (Figure 1). The drop in hotel-related services was sharpest in countries that depend heavily on tourism, such as Egypt and Jordan, where international tourist arrivals dropped dramatically and have yet to recover. Because agricultural production and food systems have been relatively more resilient than other sectors in most MENA countries, the agrifood system offers a strong foundation for post-pandemic recovery and transformation.

SHARPER INCOME DROP IN WEALTHIER, URBAN HOUSEHOLDS; GREATER HARDSHIP FOR POOR

Overall, household incomes during the main lockdown period from April to June 2020 fell by 8 percent in Egypt, 22 percent in Jordan, and 15 percent in Sudan (Figure 2). Although most households were hurt by the economic slowdown and the initial fall in remittances, urban households experienced greater absolute losses than rural households. Urban dwellers are primarily employed in the industrial and service sectors – including trade, transport, hotels, and gyms and fitness services – which were disproportionately affected by lockdowns and mobility restrictions. Rural households have also lost income, though less in absolute terms than their urban counterparts. In Yemen, where the sharp fall in remittances is believed to be the main driver of impact, the poor have been hit harder than the better-off, with an estimated income loss of 21 percent, compared with 12 percent for wealthier households.[10] Whether rural or urban, poor households are likely to find it harder than wealthier households to cope with income losses. These drops in household incomes as well as other direct

FIGURE 2 Change in average household incomes during main lockdown period, 2020

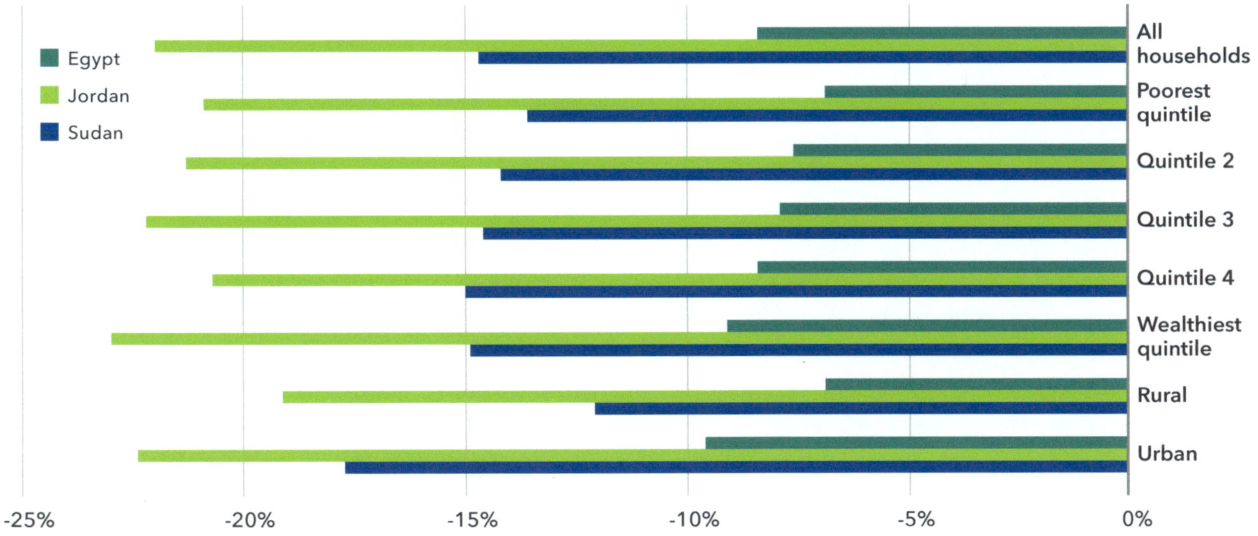

Source: IFPRI SAM multiplier models, C. Breisinger et al., *Impact of COVID-19 on the Egyptian Economy: Economic Sectors, Jobs, and Households*, MENA Policy Note 6 (Cairo: IFPRI, 2020); M. Raouf, D. Elsabbagh, and M. Wiebelt, *Impact of COVID-19 on the Jordanian Economy: Economic Sectors, Food Systems, and Households*, MENA Policy Note 9 (Cairo: IFPRI, 2020).

Note: Change is compared with same period in 2019 for each country.

public health impacts of the pandemic are also likely to adversely affect nutritional outcomes in the region.[11]

PUBLIC INVESTMENTS, SAFETY NETS, AND FISCAL LIMITS

As MENA economies and sectors begin to show signs of economic recovery, some are reviving more quickly than others, largely because of national policy responses. Although these public investments and increases in cash transfers can help economies and households recover, their fiscal implications remain uncertain. For example, Egypt's emergency response package of EGP 100 billion (US$6.4 billion) is estimated to have reduced the drop in GDP between April and June 2020 from 8.7 percent to just 1.1 percent. This rescue package is supporting the most affected sectors, including exports, tourism, and real estate. In addition, the government relaxed credit limits for businesses, provided tax-based incentives, and expanded cash transfer programs. Egypt is now implementing a public investment plan with funding of about EGP 280 billion (about $18 billion). With this government support, Egypt's economic performance improved in the second half of 2020.[12] In Sudan, workers received unemployment benefits and small businesses received three-month tax breaks. Sudan also increased spending on current social protection programs, providing SDG 12 billion (about $32 million)[13] in cash transfers to the poorest households over a period of three months and allocating SDG 10 billion for the country's essential commodity production support program, known as Selaaty ("my commodity"). In Jordan, allowances were increased for poor households, and a temporary cash transfer program was established for the unemployed and self-employed, with about JOD 81 million (about $114 million) in income support designated for seasonal workers.

Three important caveats are worth highlighting regarding the impacts of these interventions. First, the effectiveness of these policy responses depends greatly on whether the programs are well-targeted and informed by evidence on the breadth of harm suffered by each sector. Second, the level of social protection required to fully offset poor households' income losses is likely to be prohibitively expensive, especially given falling revenues from reduced economic activity and declining fiscal resources. Third, the fiscal burden and sustainability of social safety nets is a particular concern and its impact on long-term economic stability remains to be seen.

OUTLOOK FOR POST-COVID-19 RECOVERY

Agriculture and agrifood systems are playing a stabilizing role in the MENA region during the COVID-19 crisis, and these systems remain well positioned to support post-pandemic economic recovery. However, the role and post-pandemic potential of the agrifood system will depend on the successful containment of the virus in the region and globally; on government investments; and on the prevailing policy, political, and business environments. Political transitions and conflicts in some of the MENA countries, notably, Libya, Sudan, Syria, and Yemen, will likely dominate post-pandemic recovery. Similarly, smart, future-oriented public investments will be key for raising the potential of the agrifood system to support recovery and the longer-term transformation of economies. Recent and ongoing studies show that demand for digital tools has significantly increased during the pandemic.[14] Indeed, those sectors that are digitalized, like food deliveries, have been less affected by the crisis. This shift offers an opportunity to accelerate digitalization to reduce the vulnerability of the food system in the future. Increasing public and private investments in digitalization can be transformative – from providing farmers with better information through digital extension services to link smallholder farmers to markets, to better policymaking using digital tools like the recently released Agricultural Investment Data Analyzer (AIDA).[15] In the process, emphasis should be placed on promoting digital inclusion for low-skilled and illiterate people in the region. The pandemic is also a strong reminder for countries to rethink their agricultural investment priorities to include (climate) resilience, nutrition, and environmental aspects; diversify food imports and exports; and improve the business climate to allow farmers, food processors, and traders to prosper and grow.

MIDDLE EAST AND NORTH AFRICA COVID-19 TIMELINE, MARCH 2020–JANUARY 2021

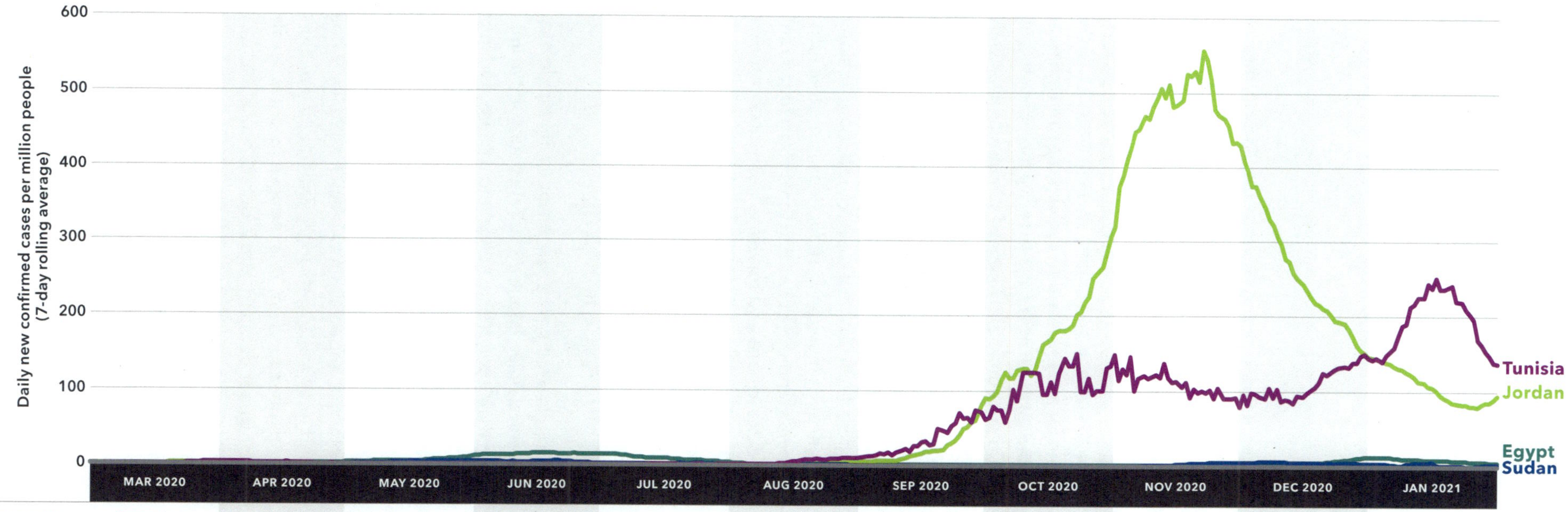

LOCKDOWNS AND HEALTH MEASURES

- **Egypt** bans public gatherings and flights, suspends school and worship activities
- **Jordan** enforces curfews and mobility restrictions
- **Sudan** introduces curfews, closes land borders and schools
- **Tunisia** imposes travel restrictions, closes schools, religious institutions, and markets
- **Sudan**'s transitional government increases health budget
- **Egypt** receives vaccine from China, opens online registration

STIMULUS

- **Egypt** allocates stimulus package
- **Jordan** offers low-interest loans for SMEs, receives IMF emergency support
- **Tunisia** creates investment funds and state guarantee for new credits
- **Jordan** establishes relief fund
- **Sudan** provides tax break for small businesses
- **Sudan** increases public sector wage, allocates funds for commodities sector
- **Tunisia** waives business VAT, allows loan payment deferments
- **Tunisia** approved for IMF loan

SOCIAL PROTECTION

- **Egypt** increases pensions, begins financial assistance to informal workers
- **Tunisia** allocates funds for unemployed, approves cash transfer program for vulnerable groups
- **Egypt** adds 100,000 families to Takaful and Karama program
- **Jordan** creates cash transfer program for self-employed and unemployed, increases cash transfers to poor households
- **Sudan** approves cash transfers to the poorest households
- **Jordan** allows hard-hit companies to cut employee salaries
- **Tunisia** approves cash transfers for tourism sector employees

Source: For COVID-19 data, Johns Hopkins University, CSSE (updated March 8, 2021). For policy data, IFPRI, COVID-19 Policy Response Portal (accessed March 2021) and news reports.
Note: Countries and policies included here were selected to provide an indication of the types of actions taken in the region.

CENTRAL ASIA

KAMILJON AKRAMOV, ROMAN ROMASHKIN, AND JARILKASIN ILYASOV

Kamiljon Akramov is a senior research fellow in the Development Strategy and Governance Division, International Food Policy Research Institute (IFPRI), Washington, DC. **Roman Romashkin** is deputy director of the Eurasian Center for Food Security (ECFS), Lomonosov Moscow State University, Moscow. **Jarilkasin Ilyasov** is a senior research analyst in the Development Strategy and Governance Division, IFPRI, Washington, DC.

Central Asia faced unprecedented challenges in 2020 as the global pandemic compounded existing problems facing the region. These included the growing effects of climate change, unstable commodity markets, and a heavy reliance on remittances and undiversified trade flows. Unlike other recent shocks, COVID-19 and the associated policy responses forced Central Asians to change daily practices and routines, many essential to their livelihoods, and disrupted connections at the local, regional, and global levels. The short-term response by governments helped to buffer the shock, while the agriculture sector proved fairly resilient. Recovery in the main destination countries for Central Asia's migrant laborers has already helped restore flows of remittances. For long-term transformation of the region's food system, however, the shock has highlighted weakness in social safety nets and digital connectivity, as well as the risk of relying heavily on remittances and a limited set of exports.

POLICY RESPONSES

The pandemic policy responses implemented by Central Asian governments appear to have been timely and appropriate. Governments pursued a mix of policies to address the health and economic needs of the population and, at the same time, to stimulate business activities in an effort to limit harm caused by pandemic restrictions. These policy responses included (1) measures to contain the spread of the virus, such as strict limits on population movement and public gatherings, restrictions on domestic and foreign travel, and lockdowns; (2) measures to mitigate impacts on household welfare and food security, such as wage and unemployment support, tax waivers for individuals, and social protection measures; and (3) fiscal measures to revive the economy, including economic and financial stimulus and tax waivers for businesses.[1] The overall cost of these fiscal revenue and expenditure measures ranged between 3 and 5 percent of GDP across countries. Monetary authorities in Tajikistan and Uzbekistan also provided macro-level financial support, including liquidity injections into banks, reductions in reserve requirements, and direct funding for lending to the real sector (goods and services).[2] Several countries, among them Kazakhstan and Russia, also implemented trade restrictions and price controls, discussed below.

ECONOMIC, TRADE, AND REMITTANCE IMPACTS

Despite swift policy responses to the pandemic, Central Asian countries suffered substantial impacts on their economies, household welfare, and food and nutrition security. The pattern of impacts followed a similar course across the region. Initially, government-mandated lockdowns and other restrictions caused a contraction of economic and business activities, especially in tourism, hospitality, wholesale and retail trade, passenger and freight transportation, and other services. This reduced incomes, weakened consumer demand, and reduced household welfare and food security. In Uzbekistan, for example, around 85 percent of small and medium enterprises (SMEs) were forced to shut down by the end of March, and in the following months unemployment rose rapidly.[3] The share of Uzbekistan's households with at least one member actively working dropped from 84 percent to 43 percent. By early June, the situation had improved somewhat, with 78 percent of SMEs open for business.[4] Similar trends were seen in other Central Asian countries.

At the macro level, Kazakhstan and Kyrgyzstan reported contractions in GDP of 2.8 percent and 8.1 percent, respectively, while Uzbekistan reported growth of only 0.4 percent.[5] The drop in growth rates was precipitated by large declines in industrial production and international and domestic trade, as well as services. Agriculture, however, recorded relatively robust growth rates despite the limited availability of and access to imported farm inputs (feed, fertilizers, and pesticides) during the spring sowing season, which was caused by higher import prices stemming from currency devaluations, logistical difficulties at the border, and a rapid decrease of inventories.[6] In January–September 2020, agricultural output grew by 5.1 percent in Kazakhstan, 2.7 percent in Kyrgyzstan, and 3.4 percent in Uzbekistan.[7] Tajikistan's overall economy, in contrast, grew at a relatively robust rate of 4.2 percent in the first three quarters of 2020; this resilience is credited to timely government support to the economy and intensified land use in agriculture.[8]

The pandemic also exacerbated structural vulnerabilities in the region's economies, especially exposure to commodity price volatility (most notably in energy prices) and heavy dependence on a few commodities and trading partners. Energy exporting countries (Kazakhstan, Turkmenistan, and Uzbekistan) were hit hard by a sharp decline in oil and natural gas demand and prices. After dropping to US$23 per barrel in April from US$66 in December 2019, oil prices partially recovered, rising above US$40 in June; however, at the end of 2020, oil was still trading at about 30 percent below pre-pandemic levels.[9] As a result, Kazakhstan's oil exports declined by more than 20 percent in the first three quarters of 2020, and its total export of goods declined by more than 18 percent.[10] The value of Uzbekistan's natural gas exports declined dramatically – by more than 70 percent – but its total value of exports of goods rose slightly because of higher gold prices, with increased gold exports more than compensating for the decline in energy exports.[11] Around the region, the negative impact on exports of services, primarily tourism and transportation, was severe – exports of services from Kazakhstan fell 34.3 percent; from Kyrgyzstan, 51.8 percent; and from Uzbekistan, 42.6 percent.[12]

Intraregional trade has contributed to mitigation of the pandemic's adverse effects on food and nutrition security in the region. For example, Kyrgyzstan increased agrifood exports by 25 percent to markets in the Commonwealth of Independent States (CIS), including Russia. However, Kazakhstan initially banned the export of wheat and wheat flour, later replacing the ban with export quotas. Kazakhstan also introduced a six-month ban on export of live animals, and Kyrgyzstan followed suit in late 2020. These restrictions raised food security concerns in food-importing Central Asian countries, some of which rely on Kazakhstan for half or more of their imported calories. In Uzbekistan, the amplified food security concerns led policymakers to put a brake on wheat market liberalization reform and to maintain the country's wheat self-sufficiency policy and price regulations. Such policies impede the allocation of additional arable land for high-value horticulture crops, like fruits and vegetables, in which Uzbekistan has a comparative advantage and which could support rural transformation.[13]

The pandemic-related disruption of remittance flows initially put additional pressure on Central Asian economies. Kyrgyzstan and Tajikistan, in particular, were hit by declining remittances from the Russian Federation. When the Russian economy suffered the double blow of the pandemic and falling oil prices, deterioration of its labor market and depreciation of the Russian ruble reduced opportunities and wages for migrant labor from Central Asia. Our analysis suggests that total monthly remittances from Russia to Central Asia dipped substantially from March to May 2020 but recovered in June. Similarly, the average amount of individual remittance transfers from Russia to Central Asian countries through the payment systems fell in March 2020 to about 50 percent of 2019 levels for that month, but by June reached nearly 70 percent of 2019 levels (Figure 1).

At the country level, remittances from the Russian Federation to Kyrgyzstan, Tajikistan, and Uzbekistan declined by 23.6 percent, 38.7 percent, and 13.1 percent, respectively, in the first half of 2020, compared with the same period in 2019. It is worth noting that the pandemic-related declines in remittances to Tajikistan

FIGURE 1 Remittance flows from the Russian Federation to CIS countries through payment systems, 2019–2020.

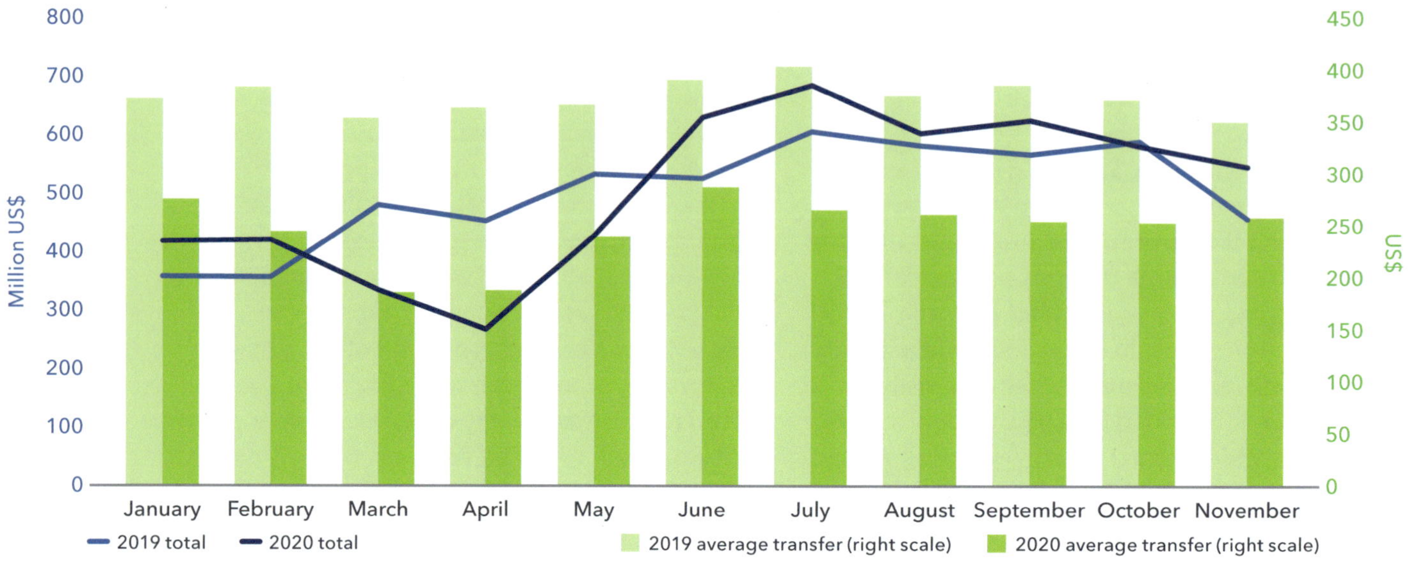

Source: Based on data from the Central Bank of the Russian Federation.

Note: CIS = Commonwealth of Independent States.

and Kyrgyzstan are comparable to those resulting from the 2015 Russian financial crisis (caused by sharp declines in energy prices). In contrast, remittances to Uzbekistan appear to be more resilient now compared with the previous crisis.

Evidence suggests that, globally, countries with robust digital connectivity mitigated up to half of the negative economic impacts of the pandemic by shifting education, healthcare, and public services as well as retail trade to online platforms.[14] However, digital connectivity in Central Asian countries is generally poor, especially in rural areas. This isolation restricted economic and social opportunities during the lockdown. As a result, many households and individuals in the region missed out on employment opportunities and could not access quality education, healthcare, or other public services.[15]

IMPACT ON HOUSEHOLDS

The pandemic has had a significant negative impact on household welfare, food security, and nutrition in Central Asia. A household phone survey conducted by IFPRI in rural areas of southern Tajikistan provides evidence of these impacts, which are likely to be similar to rural areas of other countries in the region.[16] Incomes have fallen for more than 40 percent of households, including both the poor and non-poor. Job losses have affected almost 20 percent of households, and even those who still have jobs face numerous workplace challenges.

Migration for work remains central to livelihoods in this region, despite COVID-19. Nearly half of surveyed households in Tajikistan had at least one migrant laborer before the pandemic. About 8 percent of these migrants returned home because of the crisis. Of the approximately 45 percent of households that still had at least one migrant laborer abroad, nearly 80 percent reported reduced remittances.

The pandemic shock reduced household spending, especially by the poor and on food. As household incomes fell, poor households depleted their savings; and the less-poor, who generally own more, depleted their assets. These patterns played out across the region. For example, a community survey conducted by the United Nations Development Program in Uzbekistan found that the pandemic led to reduced household incomes, depleted savings, less-diversified food consumption, and job losses. These outcomes apply to both rural and urban communities.[17]

The IFPRI phone survey also provides evidence of the impact on diets and dietary diversity. Many households switched from growing high-value crops (vegetables) back to staple food crops (cereals, mainly wheat), most likely because of food security concerns. For example, the share of households in Tajikistan cultivating cereals increased from 2018 levels across household subsistence plots (from 8 to 24 percent), "presidential" (subsidiary) plots (from 44 to 62 percent), and *dehkan* (commercial) farm plots (from 24 to 55 percent).[18]

Social safety nets to address poverty and food insecurity in the region are weak, fragmented, and poorly targeted. Less than 20 percent of the population in the lowest income quintile is covered by social assistance in Kyrgyzstan and Tajikistan. In Kazakhstan, the situation is somewhat better, with about 40 percent of the poorest population receiving social support from the government.[19] During the pandemic, Central Asian governments have increased social protection, primarily through a combination of direct cash transfers (mainly Kazakhstan) and in-kind support (all countries), but levels of support remain low.

LOOKING FORWARD

As of early 2021, Central Asia's main trading partners, China and Russia, appear to have begun a broad economic recovery, which – barring further misfortune – will likely help the region's other economies by reviving trade, investment, and remittances. Nevertheless, Central Asia will continue to face uncertainties and shocks. Recovery in services, particularly in the people-intensive tourism sector, will take time. Moreover, intraregional trade continues to face significant institutional and policy hurdles related to nontariff barriers to trade, sanitary and phytosanitary regulations, and public and private food safety and quality standards. Over the past two decades, Central Asia established relatively stable regional agrifood value chains. However, the pandemic disrupted these trade relations, as export bans and quotas for agrifood products raised food security concerns in neighboring countries. Achieving long-term regional food and nutrition security will require free and stable trade in food products, especially during crises such as the current pandemic.

Central Asian countries have also reduced poverty significantly in recent years. However, the pandemic may have pushed an additional 1.5 to 1.9 million people in the region below the poverty line in 2020.[20] Inequality may also increase, as individuals employed in informal sectors and services and households relying on remittances continue to suffer disproportionately from the adverse effects of the pandemic. The region needs more secure, well-established social safety nets that address people's needs without exceeding public sector resources.

The pandemic has also exposed the need for structural reforms to develop a competitive, business-friendly environment and promote labor mobility within countries and across the region as well as into more productive formal sectors. Likewise, the crisis has highlighted Central Asia's currently low rates of digital penetration and connectivity. Investments in information and communications infrastructure and digital technologies will be integral to long-term recovery in the region and will help it to expand its digital economy and accelerate the implementation of modern technologies, such as precision agriculture and unified digital market platforms. These have the potential to provide enormous benefits for agrifood value chains and food system transformation.

CENTRAL ASIA COVID-19 TIMELINE, FEBRUARY 2020–JANUARY 2021

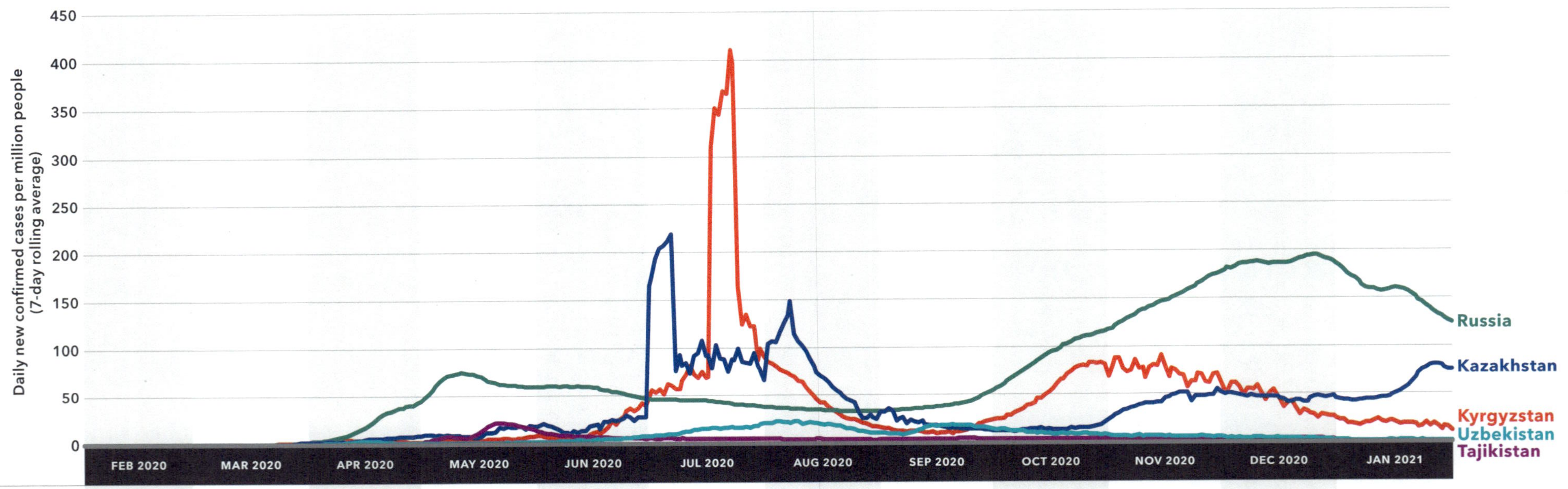

LOCKDOWNS AND HEALTH MEASURES

- All countries impose travel restrictions for China
- **Russia** and **Uzbekistan** suspend international travel
- **Kazakhstan** and **Kyrgyzstan** declare state of emergency
- Partial lockdowns in most countries, strict lockdowns in some cities
- **Tajikistan** restricts public gatherings
- **Uzbekistan** imposes strict lockdown plus stay-at-home campaign, increases funding for healthcare and quarantines

STIMULUS

- **Kazakhstan**, **Uzbekistan**, and **Kyrgyzstan** offer tax waivers and extensions
- **Uzbekistan** provides interest subsidies
- **Kyrgyzstan** adopts fiscal package to support business recovery
- **Tajikistan** announces fiscal measures and tax waivers to support businesses
- **Russia** supports temporary job creation for the unemployed

SOCIAL PROTECTION

- **Russia** announces compensation fund for quarantined citizens and businesses
- **Kazakhstan** provides monthly compensation to unemployed
- **Tajikistan** institutes payments to vulnerable families, suspends utility fees
- **Uzbekistan** increases funds to support low-income families and families with children

FOOD SECTOR INTERVENTIONS

- **Russia** subsidizes flour producers (beginning Dec. 2019)
- Most countries begin close monitoring of food prices and build emergency food supplies
- **Kazakhstan** imposes price caps and reduces VAT on food
- Most countries impose trade restrictions, including export bans, on staple crops and livestock
- **Russia** and **Kazakhstan** offer concessional loans to farmers
- **Uzbekistan** initiates a food aid campaign
- **Uzbekistan** provides support to horticultural exports through subsidies

Source: For COVID-19 data, Johns Hopkins University, CSSE Database (updated March 8, 2021). For policy data, IFPRI, COVID-19 Policy Response Portal (accessed March 2021) and news reports.

Note: Countries and policies included here were selected to provide an indication of the types of actions taken in the region. The Russian Federation is included because it is a crucial trade partner for Central Asian countries, and it is a major destination for regional migrant labor.

SOUTH ASIA

SHAHIDUR RASHID, AKHTER AHMED, AND ABDUL WAJID RANA

Shahidur Rashid is director for South Asia, International Food Policy Research Institute (IFPRI), New Delhi. **Akhter Ahmed** is country representative, IFPRI, Dhaka. **Abdul Wajid Rana** is program leader, Pakistan Strategy Support Program, IFPRI, Islamabad.

Prior to the pandemic, South Asia enjoyed years of economic progress, with fast growth, a rapid reduction in poverty, and improvements in many social indicators.[1] Although poverty and malnutrition rates remained high across the region, evidence of economic and food system transformation – notably, the declining share of agriculture in the region's economy, rising wages, and changes in dietary patterns – was strong.[2] The pandemic interrupted this long streak of impressive performance, and early predictions of its likely impact were dire. The region's GDP growth forecast was revised downward by 14 percent and overall unemployment was projected to increase by 28 percent, with youth unemployment perhaps reaching 72 percent.[3] Alarming predictions were also made regarding poverty and food insecurity, with extreme poverty expected to increase by as much as 50 percent, equivalent to an additional 72 million ultra-poor.[4] But the actual impact thus far has been less grim. Food systems and health systems have shown remarkable resilience, economic contraction has been less severe than expected, and the overall outlook appears more positive. Yet, pandemic-related challenges remain, and the experience has highlighted vulnerabilities that the region will need to tackle to ensure a better food system for future generations.

COVID-19 SHOCKS AND POLICY RESPONSES

PREVENTIVE MEASURES

South Asian political leadership took preventive measures seriously. Lockdowns were initiated in all countries in the first days of the pandemic, when case numbers remained in the hundreds and deaths near zero. Common measures included closing academic institutions, bans on public gatherings, suspension of sporting events, and travel bans. Initially announced for three weeks, strict lockdowns were either extended or later reimposed in almost all South Asian countries, lasting from 3 weeks in Afghanistan up to 17 weeks in Nepal.[5] The targeted lockdowns that followed – such as quarantining specific areas or communities in cities and restricting large gatherings and restaurant occupancy – have generally been much longer. Both the planning and enforcement of initial lockdowns varied across the region. For example, India imposed a strict travel ban, leaving many migrant workers stranded; but Bangladesh declared its lockdown a "general holiday" without a travel ban, and 10 million city-dwellers were able to return to their villages. In Pakistan, however, the prime minister deemed a complete lockdown unfeasible, given the country's level of poverty and the nature of livelihoods.

Impacts of the lockdowns in several countries appear to be reflected in infection rate trends (Figure 1). This is particularly true for the Maldives, which suffered a major spike in infections after lifting travel restrictions in July 2020 and another following the rebound in tourism during the Christmas holiday. Tourism likely played a role in Nepal as well, as its first spike correlates with the decision to reopen for trekking and other tourism.

Regionwide, the initial lockdowns kept both infections and COVID-19-related deaths low, but a spike in late summer brought the number of cases up to 1,500 per million in September. But new measures – such as targeted quarantines and strict enforcement of mask-wearing – brought the numbers down quickly. As of January 2021, cumulative cases had reached 12 million and related deaths 175,000. These are impressive numbers for the most densely populated developing region.

FIGURE 1 COVID-19 cases in South Asia

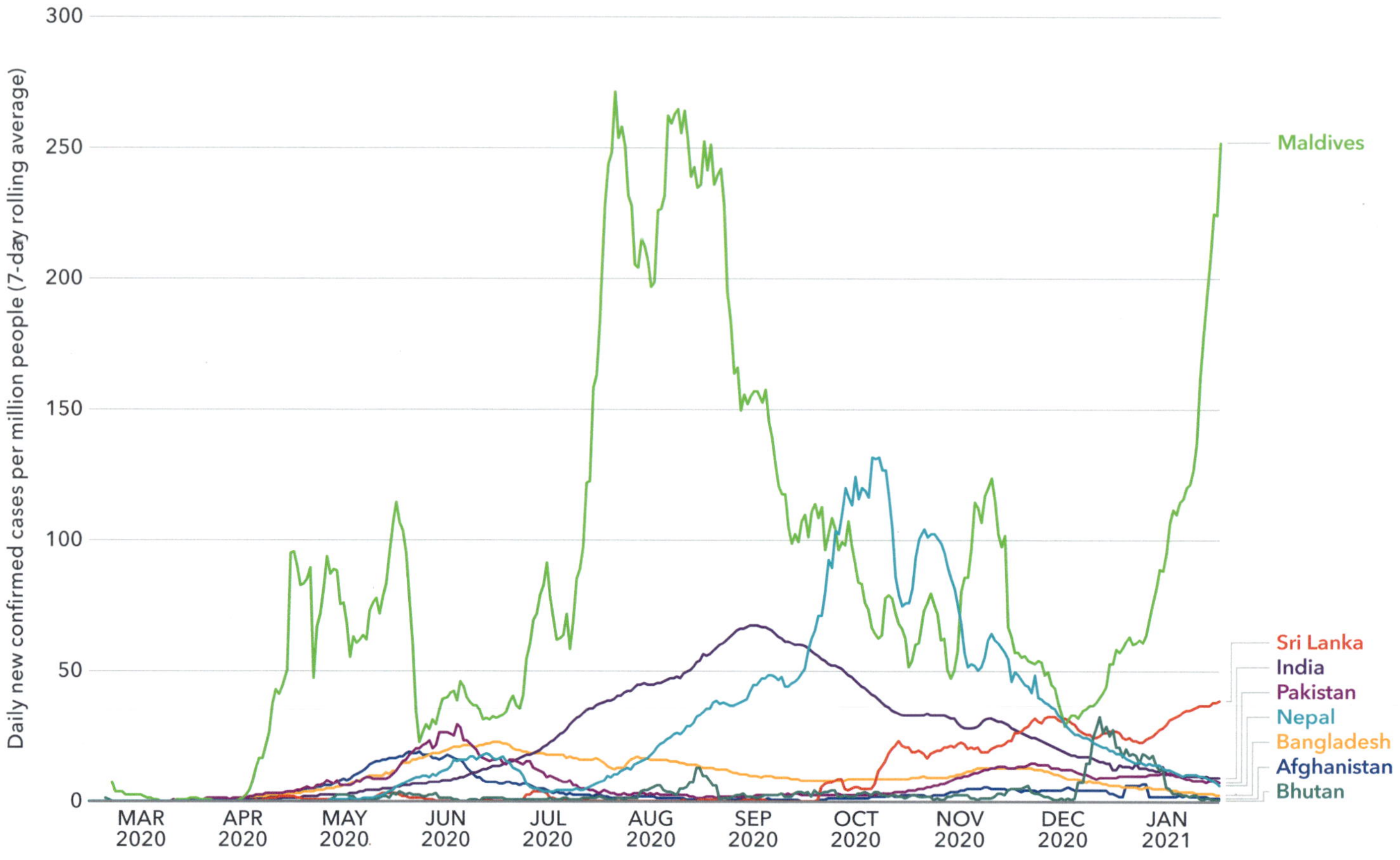

Source: Johns Hopkins University, CSSE COVID-19 Data (updated March 11, 2021).

POLICY ACTIONS TO MITIGATE IMPACTS

With low case numbers, many experts considered the region's initial lockdowns to be a bigger shock to the economy than the pandemic itself.[6] Moreover, enforcement of the lockdowns was challenging in part because they coincided with the winter harvest and spring planting, and with several major religious festivals.[7] These celebrations were overshadowed by images of the lockdowns' impacts – such as thousands of migrant workers walking on India's empty highways and normally bustling cities at a standstill. These images shaped the region's policy responses – early government actions involved a wide range of policies to save lives, protect livelihoods, and stimulate economies.

Key policy responses in the region, including macroeconomic policy responses, have been tracked by several institutions and development partners, including IFPRI, the Asian Development Bank, and the World Bank. Here, we focus on the policies related to agriculture, social protection, agricultural pricing, and labor markets.

From the onset of the pandemic, experts agreed that the region's top policy priorities should be scaling up social protection, supporting agriculture, and maintaining food price stability. These priorities were clearly reflected in the programs initiated by Bangladesh, India, and Pakistan. Our initial estimates of the allocation of public funds point to three patterns in regional spending (Figure 2). First, spending in response to the crisis has been high, ranging from 1.4 percent of GDP (about US$1.2 billion) in Sri Lanka to over 12 percent (about $287 billion) in India. Second, except for Afghanistan, the overwhelming share of government expenditure was for monetary and fiscal policy measures, suggesting that while initial emphasis was on protecting the poor and stabilizing food prices, the focus later shifted to stabilizing

FIGURE 2 Budget allocation by pandemic policy response categories in South Asia

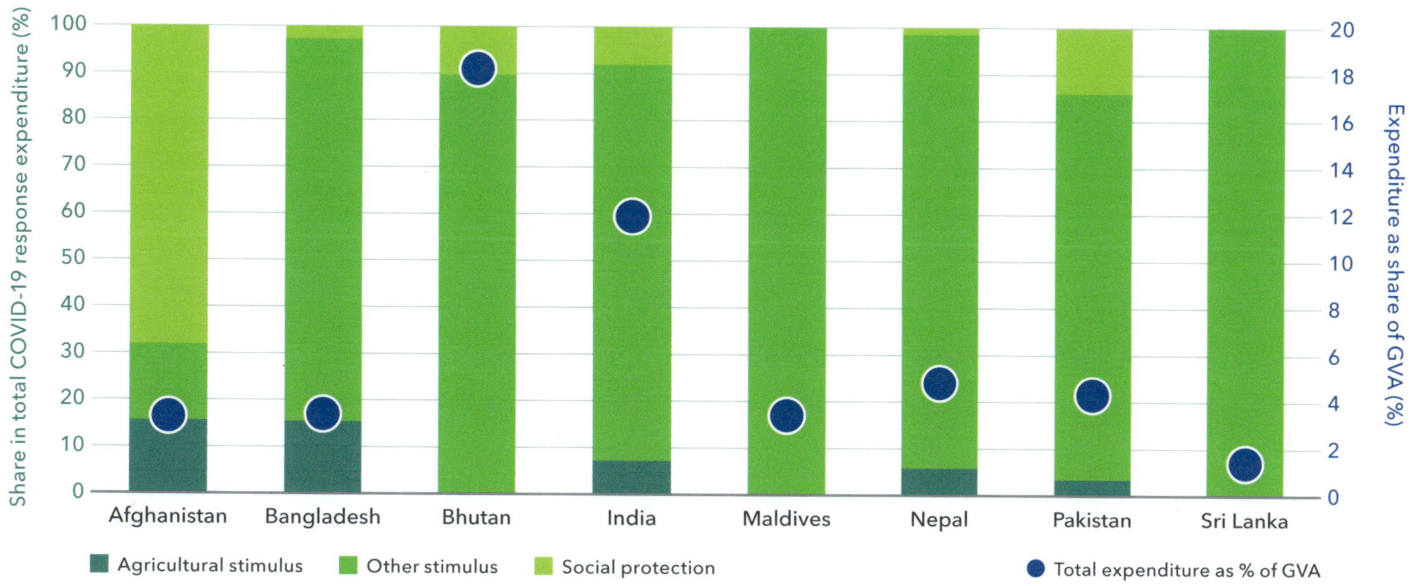

Source: Bangladesh: Bureau of Statistics; India: Press Note on First Advance Estimates of National Income 2020/21, Ministry of Statistics and Programme Implementation; Nepal: Central Bureau of Statistics; Pakistan: Bureau of Statistics and State Bank of Pakistan; Bhutan, Maldives, and Sri Lanka: World Economic Outlook, IMF, October 2020; COVID-19 Policy Response Portal, IFPRI; Policy Responses to COVID-19, IMF.

Note: GVA = gross value added. The shares of expenditure on safety nets and agriculture represent only the costs of scaling up the existing programs.

economies. Third, the cost of scaling up agriculture and social protection programs accounted for more than 15 percent of the total pandemic-response budget (as shown in Figure 2), even though the region's three largest economies (India, Pakistan, and Bangladesh) already had large safety-net and agricultural support programs.[8]

RESILIENCE AND VULNERABILITIES IN SOUTH ASIAN FOOD SYSTEMS

EVIDENCE OF RESILIENCE

The pandemic has highlighted both resilience and vulnerabilities in the transformation of food systems in South Asia. Recent estimates of food price stability, agricultural growth, wage rates, unemployment, and poverty rates have proved most of the dire predictions wrong and suggest that the region's policy actions are paying off.[9] Agriculture sectors have registered growth in almost all countries, in part because of the exemption of agriculture from lockdown restrictions and effective use of the existing safety-net and agricultural development infrastructure in responding to the crisis.

In terms of GDP growth, all countries performed better than the forecasts (Figure 3).[10] Bangladesh achieved 5.4 percent real GDP growth, and Nepal achieved 2.4 percent growth. But the region's two largest economies, India and Pakistan, experienced negative growth of −7.2 percent and −0.4 percent, respectively, enough to cause an overall contraction of 5.4 percent in the regional economy.

Recent surveys conducted by IFPRI and its partners suggest that public transfer systems for food security, health, and nutrition have worked well in the region's large countries. In India, two recent studies show that the initial support package, called Pradhan Mantri Garib Kalyan Yojana (PMGKY), was timely and effective in reaching smallholders.[11] Similarly, surveys by IFPRI and national partners in seven Indian states indicate that disrupted health services have been restored and are adapting to the new challenges. One of the key conclusions of the study is that India's rapid policy actions and effective coordination across national, state, and local institutions helped buffer the initial shocks to health and nutrition programs.[12] This success reflects India's decades of investments in social-safety-net infrastructure, particularly recent investments in direct and cash benefit transfers (see Chapter 5).[13]

FIGURE 3 Growth in real GDP in 2019 and 2020, and 2020 forecast

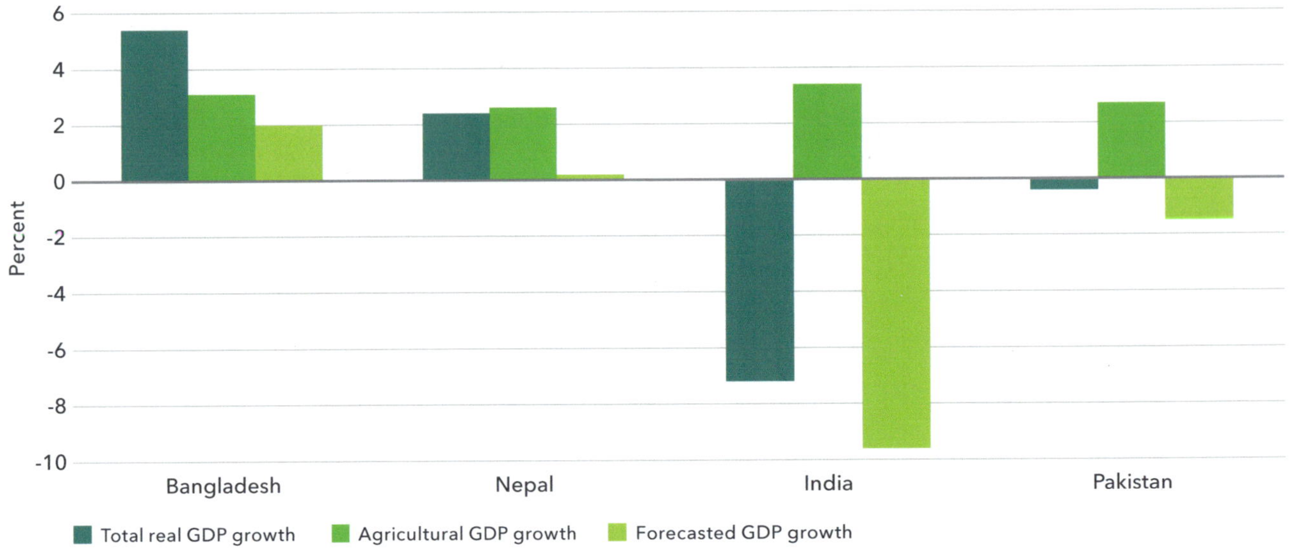

Source: Bangladesh: Bureau of Statistics; India: Press Note on First Advance Estimates of National Income 2020/21, Ministry of Statistics and Programme Implementation; Nepal: Central Bureau of Statistics; Pakistan: Bureau of Statistics and State Bank of Pakistan.

Note: Regarding the estimates, fiscal years for Bangladesh, Nepal, and Pakistan are July–June; April–March for India.

Pakistan's Ehsaas program, which provides direct cash transfers, targeted 12 million households (about 80 million people) with a budget of $900 million. Coverage was extended to an additional 6 million families during the pandemic, suggesting that almost half of Pakistan's population was covered by these programs; such rapid expansion was possible because of the government's earlier investments in building the necessary infrastructure.

Bangladesh also has several social protection programs that provide cash and food transfers to vulnerable populations. In response to the pandemic, the government increased the social protection budget from 2.9 percent of national GDP in 2019/20 to 3.01 percent in 2020/21 (reaching about $11.24 billion). In addition, as an Eid-ul Fitr gift from the prime minister, the government made a one-time cash transfer of about BDT 2,500 to each of five million poor families, totaling BDT 12.5 billion (roughly $150 million). This expansion of transfers was made possible by earlier government investments in mobile financial services, including digitalization of social transfers.[14]

Recent regional poverty estimates suggest that safety nets and other social transfer programs have largely performed well. In Bangladesh, over three-quarters of the vulnerable nonpoor are estimated to have fallen below the poverty line in June–July 2020, largely because of rising unemployment.[15] But the country has recovered from the initial shock, making it likely that the actual poverty rate is lower than the government's official third-quarter estimate of 29.5 percent (still up 9 percent over 2019). Moreover, the most recent round of an IFPRI-Cornell phone survey (completed in January 2021) suggests that many food security indicators improved after drastically worsening in the pandemic's early months (when shares of food-secure households plummeted from 53 percent to 12 percent in rural areas and 65 percent to 8 percent in urban areas), and are now stronger than they were in 2019.[16]

Pakistan's poverty rate was initially projected to increase from 27 percent to 43 percent, but things gradually improved after the economy was re-opened. By the end of 2020, the estimated poverty rate was only about a percentage point higher than the pre-pandemic rate.[17] Surveys conducted to assess the pandemic's impact at the end of 2020 found that 90 to 95 percent of Pakistani workers affected by the initial shock had recovered; and 36 percent of households felt the pandemic had had a serious financial impact.[18]

RISKS AND VULNERABILITIES

The pandemic has highlighted several vulnerabilities of South Asian food systems. Perhaps most important is the vulnerability of South Asian labor markets, especially in the nonfarm and informal sectors, which are central to economic and food systems transformation. These sectors were decimated by the strict lockdowns. Unable to work, migrant workers returned to their home villages, creating labor scarcities in regions with agricultural surpluses that depend on these informal workers. In agricultural deficit regions, food insecurity increased for wage workers due to low wages, job losses, and reduced incomes. While there are encouraging signs of recovery, some impacts may linger in the long run.

The pandemic-related short-term shocks to labor markets may also have long-term consequences in terms of nutrition, poverty, and overall wellbeing, particularly for many South Asian children.[19] As in other developing regions, nutritious diets are not affordable for the poor in South Asia.[20] In India, for example, almost half of all poor people cannot afford a nutritious diet.[21] With massive unemployment in informal sectors, poor households have suffered a reduction in diet quality that will likely have long-term impacts on productivity.[22] In addition, the majority of children were deprived of formal schooling during the pandemic because they did not have Internet access. Unless appropriate policy actions are taken, the pandemic's lasting impacts will exacerbate inequality, reduce lifetime earnings, and limit the ability of these children to escape poverty in the future.

South Asia's international remittance inflows also proved to be vulnerable. More than 11.5 million South Asians, mostly unskilled laborers, work abroad and in 2019 remitted an estimated $108 billion.[23] As of December 2020, remittance flows had declined by about $10 billion for the region.[24] For South Asia's immigrant workers, who are generally poor and are often the main breadwinners for their families, this has meant a significant loss of income. The drop in remittances also has macroeconomic implications, as these flows are an important source of foreign exchange for some countries, accounting for as much as 28 percent of GDP in Nepal and 8 percent in Pakistan.

While these new challenges must be tackled, policymakers should not lose sight of the old challenges as the region seeks to promote an inclusive and sustainable food system transformation. South Asia faces continuing and, in some cases, intensifying problems related to climate change, natural disasters, poor food safety, and distortionary policies. The fall armyworm devastated Afghanistan, Pakistan, and parts of India and Nepal in 2020; Bangladesh struggled with flooding during the pandemic; and in India, bird flu caused a nationwide food-safety scare. Distortionary policies, and increasing costs of implementing them, remain despite overwhelming evidence of their negative impacts and of the potential to repurpose these much-needed resources for climate-smart investments or to build robust food-safety institutions. For example, public food transfer programs linked with agricultural price policies are fraught with leakage and inefficiencies.[25] But reforming such longstanding agricultural policies will be difficult for South Asia, as the recent farmer protests in India have shown.[26]

LOOKING FORWARD

Agriculture and food systems in South Asia have demonstrated unexpected resilience in the face of the pandemic. However, while there are success stories that can inform future policy, there is no room for complacency. Growth in the nonfarm and service sectors is an essential precondition for economic and food system transformation, and COVID-19 has reminded South Asia how vulnerable these sectors are. During the years of rapid growth, the share of populations engaged in these sectors grew rapidly, but infrastructure and institutions to protect the vulnerable against large-scale shocks did not. This was particularly evident from the plight of migrant workers and the sharp increase of food insecurity in major urban centers following the lockdowns. These vulnerabilities must be addressed. At the same time, the region must not lose sight of the structural challenges to transforming economic and food systems. Tackling those challenges will require long-term commitment. There is reason to hope that the pandemic experiences will help to catalyze a shift in policy priorities. The recently announced Indian budget – with significant increases for addressing climate change, health and nutrition, and enhanced biosecurity preparedness – suggests that change is on its way.

SOUTH ASIA COVID-19 TIMELINE, MARCH 2020–JANUARY 2021

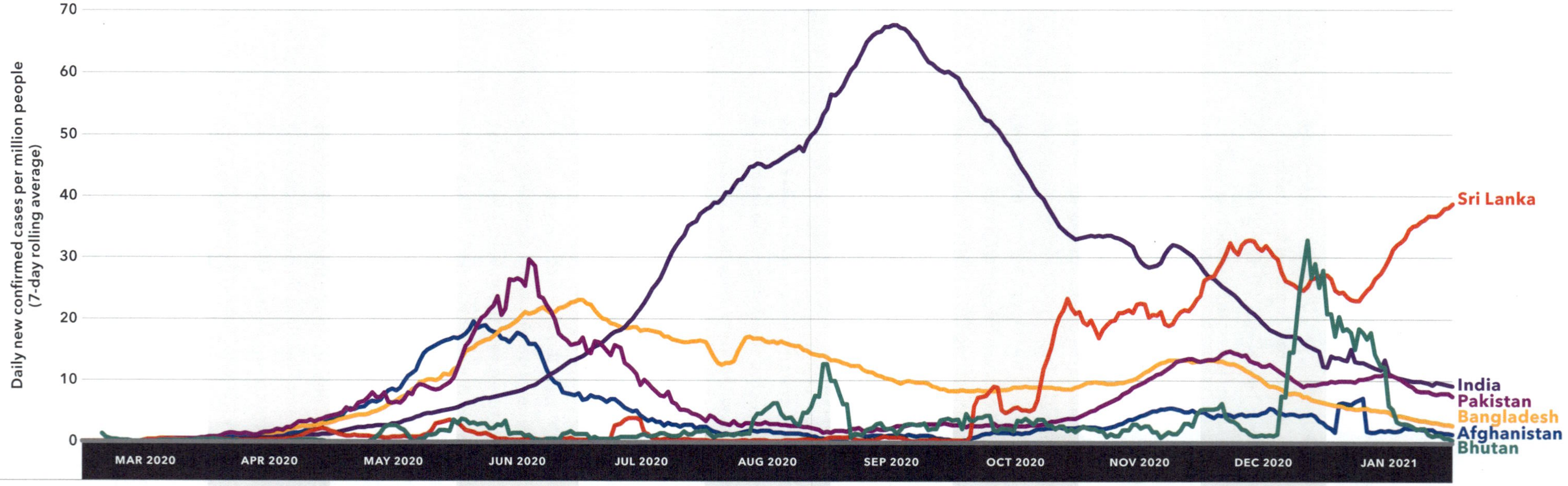

LOCKDOWNS AND HEALTH MEASURES

- **Bangladesh** announces "general holiday"
- **India** and Nepal begin lockdown
- **Pakistan** begins lockdown, exempting farm sector, groceries
- **Afghanistan** begins lockdown, in place until August
- **Bangladesh** imposes curfew, bans flights
- **Pakistan** sets up national coordination committee
- **Sri Lanka** extends work-from-home period
- **Bangladesh**, **India** extend lockdown
- **Bangladesh** ends general holiday, focuses on health precautions
- **Pakistan** announces "smart lockdown"
- Nepal ends lockdown
- **Pakistan** begins reopening
- **India** removes curfew, allows small gatherings
- **Bangladesh** removes all movement restrictions
- **Pakistan** reimposes smart lockdown
- **India** largely open

STIMULUS

- **Bangladesh** announces stimulus for exports
- **Pakistan** fiscal package supports SMEs, industry
- **Bangladesh** prepares large stimulus package
- **Pakistan** eases loan payments
- **Bangladesh** receives concessional loans for pandemic recovery
- **India** approves stimulus funds for SMEs, infrastructure
- **Bangladesh** provides cash incentives to exporters
- Nepal receives loan for agriculture sector and recovery

SOCIAL PROTECTION

- **Pakistan** expands Ehsaas cash transfers
- **India** approves funds for social protection, employment generation
- **Bangladesh** extends FFP
- **Pakistan** disburses cash to 12 million daily laborers

FOOD SECTOR INTERVENTIONS

- **Bangladesh** waives taxes on animal and fish feed
- **Pakistan** supports wheat procurement, farm inputs
- **Bangladesh** subsidizes fertilizer sector
- **India** approves funds for agriculture
- **Pakistan** allows private sector wheat imports
- **Pakistan** launches support for farm inputs

Source: For COVID-19 data, Johns Hopkins University, CSSE Database (updated March 8, 2021). For policy data, IFPRI, COVID-19 Policy Response Portal (accessed March 2021) and news reports.
Note: Countries and policies included here were selected to provide an indication of the types of actions taken in the region.

EAST AND SOUTHEAST ASIA

KEVIN CHEN AND YUE ZHAN

Kevin Chen is a senior research fellow, International Food Policy Research Institute (IFPRI), Beijing, and a chair professor, Zhejiang University, Hangzhou, China. **Yue Zhan** is a research assistant, IFPRI, Beijing.

COVID-19 is having an unprecedented impact on people's livelihoods worldwide and has triggered a severe global economic downturn. In East and Southeast Asia, however, successful containment in most countries is allowing for relatively rapid economic reopening and recovery. East Asia's continued upturn, especially China's, will be critical for recovery in the rest of the world. Despite the remarkable resilience of food systems, the effects of the pandemic on jobs and incomes in the agrifood sector have been severe and demand broad social protection programs. In addition, interregional trade and cooperation can contribute to the region's recovery. COVID-19 has been a resounding wake-up call to better prepare our food systems for future pandemics and other disasters. East and Southeast Asian countries can use the recovery as an opportunity to increase investments in reorienting food systems toward more sustainable and resilient trajectories, while simultaneously addressing inequality and accelerating productivity through the development of digitalization in the agrifood sector on a much greater scale.

LOCKDOWNS AND ECONOMIC LOSSES

In response to the COVID-19 outbreak that emerged in East Asia in late 2019, countries in the region employed more stringent containment measures than elsewhere, including mandatory lockdowns, closing of schools and government offices, and restrictions on travel, public gatherings, and religious activities.[1] On average, countries in the region declared lockdowns or a state of emergency about 17 days after confirmation of 50 cases.[2] As of late 2020, East and Southeast Asia had, by and large, outpaced other developing regions in containing the pandemic. Some countries, like China and Viet Nam, were already enjoying a revival of economic activity as of early 2021, with restrictions on mobility to stem the pandemic largely eased. Others, including Indonesia, the Philippines, and Myanmar, are still struggling to curb the virus's spread (Figure 1) and have therefore kept in place or reintroduced restrictions, which are having severe economic consequences.

Economic growth in East and Southeast Asia as a whole slowed significantly in 2020. Estimates suggest that the region's total GDP grew by just 0.9 percent[3] – with 2 percent growth in China offsetting the 3.5 percent contraction in the rest of the region (Figure 2). In 2021, regional growth is projected to accelerate to 7.4 percent.[4] East Asia's recovery, especially China's strong performance, should help fuel a global rebound.

Remittances from outside the region are estimated to have declined by 13 percent in 2020 as a result of the pandemic;[5] but the pandemic's effect on domestic remittances from urban to rural areas is less clear. With many urban workers losing jobs and moving back to rural areas, the likely decrease in domestic remittances will hurt many rural households. In Myanmar, for example, the sharp decline in both domestic and international remittances is likely to continue for at least a year, pushing many households into poverty and food insecurity.[6]

The most serious impact of the pandemic in the region has been the devastating loss of jobs and livelihoods. Poverty in developing East and Southeast Asian countries could increase for the first time in 20 years – and with it, food insecurity.[7] Poor households facing income losses will be forced to reduce food expenditures and replace expensive foods like meat and vegetables with less nutritious options.[8] Estimates from the World Food Programme suggest

FIGURE 1 Total COVID-19 cases per million people

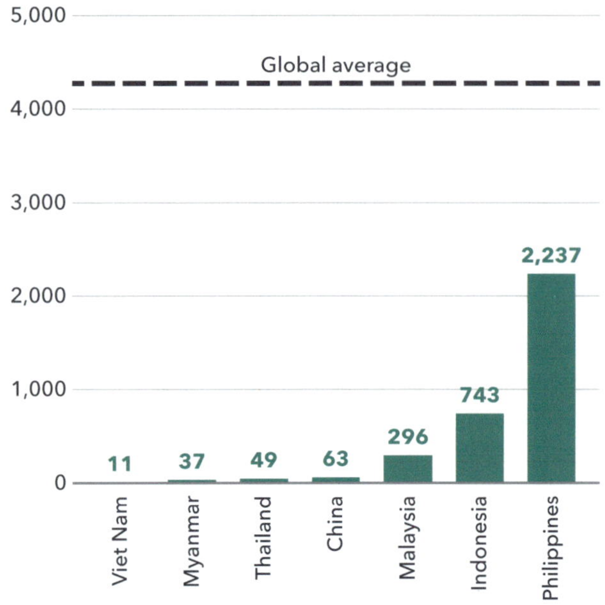

Source: Johns Hopkins University, CSSE COVID-19 Data, February 2021 update.

Note: COVID-19 cases are cumulative through early September.

FIGURE 2 GDP growth projections

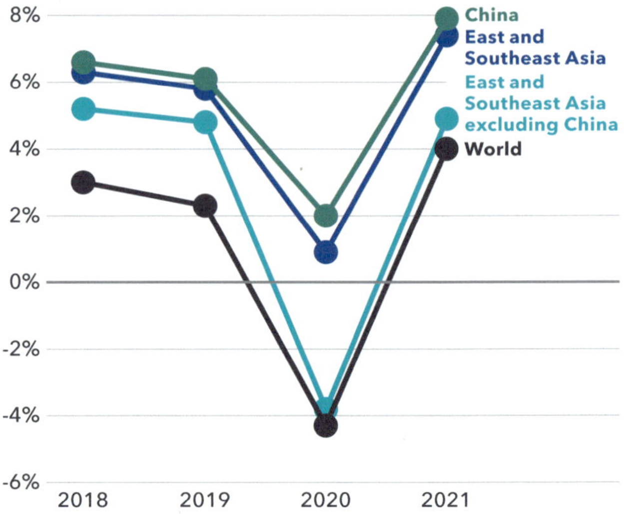

Source: World Bank, *World Bank Global Economic Prospects*, January 2021.

that the number of food insecure people in the region (excluding China) rose from roughly 93 million in February 2020 to about 112 million in June.[9] While this figure has since declined slightly – to about 108 million as of mid-September 2020 – the crisis may have lasting effects on diets and incidence of undernutrition among the poor.

To help ease these economic burdens, almost every country in the region has implemented social protection and food assistance programs, which are essential for addressing rising food insecurity.[10] Measures include food aid, cash transfer programs, support to small and medium enterprises, easier access to loans, tax breaks, subsidies, and in-kind transfers including food or food vouchers and school feeding programs. Cambodia, China, Indonesia, Mongolia, and Viet Nam have all enacted some form of exemption, reduction, or deferral of social security contributions. Several countries have also increased unemployment benefits and eased requirements for receiving these benefits. A remaining challenge is to reach the urban poor, particularly those in the informal sector who are not covered by existing social protection systems or cannot access public services. Inequality needs to be addressed in both short-term stimulus measures and long-term policy changes for food system transformation.

DISRUPTIONS TO THE FOOD SYSTEM

Agricultural production has been less affected than other sectors by the global economic downturn. Globally, food supplies remained ample in 2020.[11] Pandemic restrictions did temporarily disrupt agricultural production in some areas through labor supply shortages and farmers' reduced access to inputs, but a major impact on agricultural production is now unlikely in the region, as East Asian countries have already begun to ease movement restrictions. However, natural disasters, such as Typhoon Vongfong, which destroyed thousands of hectares of agricultural land and displaced 200,000 people in the Philippines, may affect overall crop yields.[12]

Beyond the farm, COVID-19 has affected all segments of the food system, including processing, retailers, and markets. East Asian governments have prioritized keeping food supply chains functioning and getting food to consumers. China was able to reduce food system disruptions by opening a "green channel" for agricultural inputs and food products, as well as exempting movement of food-system workers from COVID-19 lockdowns.[13] Thailand created "war rooms" – joint working groups between government departments and relevant industries – to ensure supply of specific food products such as rice, livestock, fruits and vegetables, and processed foods.[14] Delivery services for groceries and restaurant meals have expanded around the region, and in China, e-commerce companies added in-app features for contactless food delivery.[15] These delivery platforms have helped minimize the risk of infection from crowded food markets.

Evidence from China shows that, while agricultural activities are often excluded from the disease-control policies that shut down many nonfarm activities, the indirect effects of restrictions are significant because the agriculture sector has become closely integrated with the rest of the economy, especially through agrifood processing industries. Many of these have been shuttered by lockdown policies, which affects agricultural production and other food system components.[16]

Local markets throughout the region have seen price increases and volatility resulting from supply chain disruptions. For instance, retail rice prices in Lao PDR and Thailand rose about 20 percent on average in January-April 2020 compared with the same months of 2019, but rice prices fell when harvests began later in 2020. On the whole, however, food supply chains have been remarkably resilient, and domestic food prices in the region have been largely stable.[17] The significance of the pandemic for East and Southeast Asian food systems thus comes less from its impact on primary production and disruptions along supply chains than from the devastating effects on jobs and livelihoods.

REGIONAL COOPERATION AND INTRAREGIONAL TRADE

Loss of income combined with uncertainty in 2020 have drastically reduced global demand for goods and services. As East Asian economies begin to recover, intraregional trade – particularly exports to China – has increased to replace trade lost with other regions.[18] While data from China's customs agency suggest a decline in China's imports from the European Union and United States (despite the ambitious targets set by the China-US Trade Agreement signed in early 2020), there was a 7.8 percent increase in imports from the ASEAN-5 countries in the first half of 2020. Since 2019, implementation of the upgraded China-ASEAN Free Trade Area protocol has further boosted China-ASEAN trade, including agricultural trade, and ASEAN became China's largest trading partner for the first time in early 2020 (Figure 3). Recovery of food demand from China has been a boon for farmers in ASEAN countries, boosting their incomes during the economic recovery period. For example, Chinese imports of durian from both Thailand and Malaysia increased in 2020.[19]

In their initial responses to the pandemic, several East Asian countries implemented temporary export restrictions to protect domestic food supplies. For example, Viet Nam (the world's third-largest rice exporter) imposed an export ban on rice, though later replaced the ban with a less restrictive export quota. Cambodia also imposed a rice export ban from early April to late May. Myanmar temporarily suspended

FIGURE 3 Change in China's imports from the ASEAN-5, US, and EU, 2019 and 2020

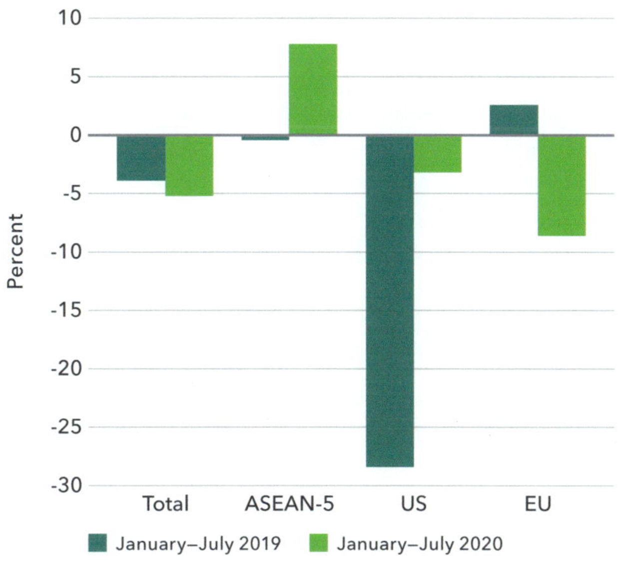

Source: World Bank, *From Containment to Recovery*, East Asia and Pacific Economic Update, October (Washington, DC: 2020).

Note: ASEAN-5 countries are Indonesia, Malaysia, the Philippines, Singapore, and Thailand.

the issuance of new export licenses from mid-March through April. Such restrictions can raise consumer prices and add to food insecurity, especially for import-dependent countries. For example, the price of Thai 5% broken rice, a benchmark global market indicator, immediately increased by about $100 per ton (more than 20 percent) when Viet Nam imposed its export ban. Other trade barriers were imposed to prevent cross-border transmission of the pandemic. For example, Indonesia and the Republic of Korea initially prohibited imports of animals and animal products from China in January and February. Such measures may remain in place for some time, with serious impacts on the global food system.[20]

Fortunately, East and Southeast Asian countries have recognized the problematic nature of food export restrictions, and most have been lifted. The ASEAN Ministers of Agriculture and Forestry released a joint statement on suitable measures to overcome pandemic-related challenges and ensure the sustainable supply of sufficient, affordable, safe, and nutritious foods that meet the dietary requirements of ASEAN populations.[21]

The signing of the Regional Comprehensive Economic Partnership (RCEP) Agreement between ASEAN countries and partners – China, Japan, Republic of Korea, Australia, and New Zealand – in November 2020 provided crucial momentum to its members' strong commitment to pursuing free trade and strengthening regional supply chains for post-pandemic economic recovery and stronger food systems.[22] The growth of regional cooperation can also provide additional food supply safety nets to address the negative impacts of the pandemic. For example, the ASEAN Plus Three Emergency Rice Reserve (APTERR) has stockpiled 787,000 tons of rice, with 87,000 tons contributed by ASEAN member countries and 700,000 tons from China, Japan, and Korea, which can effectively address short-term emergencies.[23] There are strong linkages within the region, and interdependencies among countries and food systems are no exception. Working together allows these countries to exchange lessons learned, address food system issues at the regional level, and build more resilient food systems for the future.

TOWARD AN INCLUSIVE, SUSTAINABLE, AND RESILIENT TRANSFORMATION

Even before the pandemic, there was a need for food system transformation that enhances nutrient-rich diets and public health, sustainable natural resource management, and resilience to climate change. The recovery phase offers an opportunity to address inequalities and gaps in social safety nets to ensure food security and nutrition for all. With social distancing and quarantine measures unlikely to disappear any time soon, given the potential for new outbreaks, food supply chains will require expanded use of digital innovations such as online platforms to optimize logistics and services. Longer-term responses should accelerate wider adoption of agricultural technologies such as remote sensing and geographic information systems (GIS) to address the impacts of climate change and environmental and natural resource degradation. To enhance prevention and preparedness for the multitude of shocks that can affect food systems, sustained financing is also needed – guided by data and analysis – to support improvement of risk management systems, diversification of income sources, more sustainable farming practices, and a shift away from those agricultural systems that are most vulnerable to shocks.

EAST AND SOUTHEAST ASIA COVID-19 TIMELINE, JANUARY 2020–JANUARY 2021

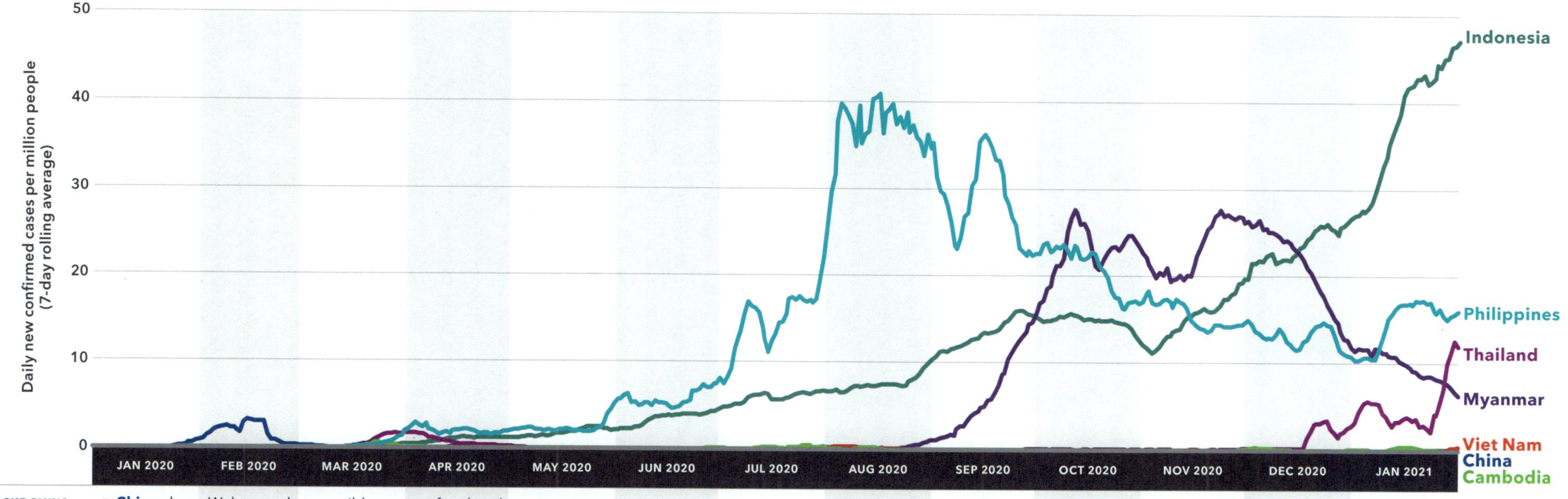

LOCKDOWNS AND HEALTH MEASURES

- **China** closes Wuhan market, possible source of outbreak
- **China** imposes travel restrictions and partial movement restrictions in Wuhan and other cities
- **Viet Nam** declares a state of emergency and bans all flights to and from China
- Joint statement from ASEAN Defense Ministers on COVID-19
- **Thailand** closes all borders and bans foreign visitors
- **Indonesia** declares a public health emergency

STIMULUS

- **China** encourages lending to SMEs and supports delay of loan payments
- **Indonesia** announces a stimulus package (US$725 million)
- **Thailand** plans for a new stimulus package worth 10% of GDP
- **Myanmar** announces IMF- and JICA-funded stimulus programs

SOCIAL PROTECTION

- **Philippines** approves program for low-income families and health workers
- **Thailand** announces payments to 8.4 million farming households
- **Cambodia** launches a cash relief program for 600,000 vulnerable families
- **Myanmar** provides cash handouts to affected households

FOOD SECTOR INTERVENTIONS

- **Indonesia** and Rep. of Korea prohibit animal and animal-product imports
- **China** opens "green channel" for the movement of agrifood products
- **Viet Nam** and **Cambodia** impose rice export bans
- **Myanmar** provides in-kind food transfers to vulnerable households
- Most countries exempt food from movement restrictions
- **Thailand** creates "war rooms" for specific food products
- Asian countries sign Regional Comprehensive Economic Partnership Agreement (RCEP)

Source: For COVID-19 data, Johns Hopkins University, CSSE Database (updated March 8, 2021). For policy data, IFPRI, COVID-19 Policy Response Portal (accessed March 2021) and new reports.
Note: Countries and policies included here were selected to provide an indication of the types of actions taken in the region.

LATIN AMERICA AND THE CARIBBEAN

EUGENIO DÍAZ-BONILLA AND VALERIA PIÑEIRO

Eugenio Díaz-Bonilla is the head of the Latin America and the Caribbean Program, International Food Policy Research Institute (IFPRI), Washington, DC. **Valeria Piñeiro** is a senior research coordinator in the Markets, Trade, and Institutions Division, IFPRI, Washington, DC.

Countries in Latin America and the Caribbean (LAC) have been hit hard by the COVID-19 pandemic. In terms of death rates per 100,000 people, 8 LAC countries, as of this writing, are among the top 20 countries in the world.[1] Since late February and early March 2020, when most of the first cases were registered, LAC governments have reacted with diverse policies, from strong lockdowns in countries such as Argentina, Chile, and Peru, to allowing economic activities to continue largely uninterrupted, as in Brazil and Mexico. However, despite substantial commitments to health interventions, social protection, and employment support, the region's GDP likely declined by about 8 percent in 2020, compared to a decline of less than 4 percent for all emerging and developing countries.[2] As a result, the pandemic will have long-term economic and nutritional impacts.

Post-pandemic, the region will need to expand support for long-term transformation of LAC food systems and redesign safety nets to address both the lingering impacts of the crisis and the vulnerabilities that underlie these impacts. Agricultural R&D systems will need to focus not only on the challenges of productivity and climate change adaptation and mitigation, but also on making food systems resilient to pandemics and other health shocks. Given the key role that LAC, the world's leading net food exporting region, plays in global food supply and global environmental sustainability and biodiversity, the region's management of the current pandemic and its aftermath will have planetary repercussions.

A combination of characteristics that distinguish the LAC region makes LAC countries particularly vulnerable to the pandemic. First, living and working conditions facilitate the spread of the virus. LAC is more urbanized than other developing regions, with about 80 percent of the population living in urban areas. This leads to more person-to-person contact. In addition, roughly half of employment is in informal activities, a proportion that is even larger for low-income groups. These activities often require in-person presence and provide no unemployment insurance. Second, already existing health conditions aggravate the impact of the virus. The rates of overweight and obesity in LAC are among the highest in the world. These conditions make people more vulnerable to the virus, both directly and through the associated noncommunicable diseases. At the same time, health systems have suffered because of the region's economic stagnation in recent years. These economic problems have also affected the vitality of LAC's democracies, reducing people's confidence in their governments, which may weaken political support for enforcement of, and compliance with, pandemic-related restrictions.[3]

Latin America's governments have implemented a variety of health measures to control the coronavirus (mostly lockdowns and closing of borders, and to a far smaller degree testing and contact tracing), while stepping up treatment efforts. Lockdowns and concerns about contagion among the population led to significant declines in mobility and economic activity. Therefore, governments also rolled out economic and social policy initiatives designed to mitigate the loss of income and employment that is affecting families and firms. To pursue these policies, countries have increased their public expenditures and injected additional liquidity into their economies through monetary policies. Yet, the size of the COVID-19 packages has differed significantly.[4] Brazil, Peru, and Chile have spent well over 5 percent of GDP through fiscal measures, plus similarly large injections of liquidity, while

FIGURE 1 Number of people who cannot afford a nutrient-adequate diet

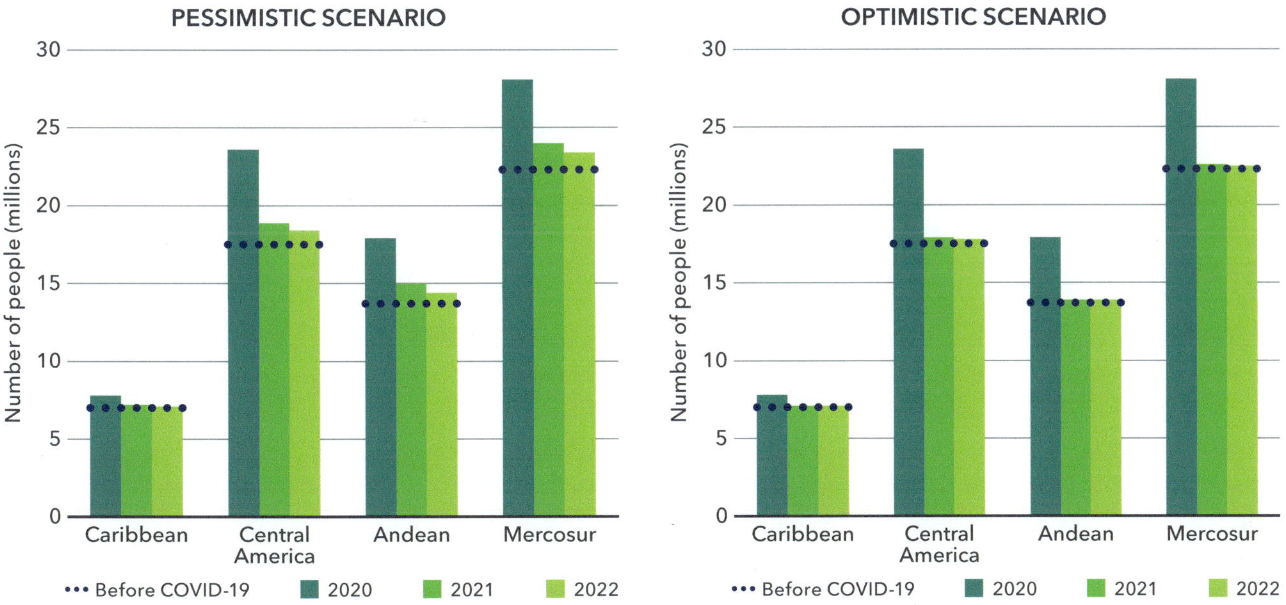

Source: E. Díaz-Bonilla, D. Laborde, and V. Piñeiro, *Covid-19: The Impact on Food Security in Latin America and the Caribbean* (Washington, DC: InterAmerican Development Bank, forthcoming).

Mexico and Colombia have spent the equivalent of 1 percent or less of GDP. Spending and monetary interventions in other countries, such as Argentina and Honduras, fall in between.

LONG-TERM IMPACTS

To explore the likely consequences of the pandemic and the policy responses over the longer term, we used IFPRI's MIRAGRODEP model to analyze some scenarios for 2021 and 2022 in LAC, focusing on poverty and nutrition.[5] Regional GDP is expected to recover from its low point during the pandemic, but will still be lower in 2022 than it would have been if the pandemic had not occurred. As a result, the number of extreme poor (those living on US$1.90 PPP per day or less) is estimated to have increased by about 11.7 million people in 2020, and will remain above 2019 levels – by somewhere between 500,000 (optimistic scenario) and 1.7 million (pessimistic scenario) people – in 2021 and 2022.

The simulations also considered the impacts on nutrition, mostly driven by incomes and wages, using different types of diets.[6] Here, we report the results for a "nutrient-adequate diet" – that is, a diet providing adequate calories plus minimum levels of all essential nutrients (Figure 1). Before the pandemic, about 60.5 million people in the region were unable to afford a nutrient-adequate diet. This number is expected to have increased by 17 million in 2020, reaching over 77 million. As the economy recovers, more people will be able to afford this diet again, but in 2022 there will still be between 800,000 and 2.8 million *more* people unable to afford an adequate diet than before the pandemic.

POLICIES FOR THE LONG-TERM TRANSFORMATION OF FOOD SYSTEMS

The long-term transformation of Latin America's food systems will need to address the vulnerabilities that have been highlighted by the pandemic, while also dealing with the structural problems existing before COVID-19. Because of the sustained negative impacts on poverty and nutrition, LAC's social safety nets and nutrition programs will need to be reevaluated. Before the pandemic, the region was spending about 1.5 percent of GDP on social protection (including a variety of safety-net and food programs), but with large differences across countries.[7] Additional expenditure on social protection will be needed, as well as adjustments in design and coverage, to address not only the poverty and food insecurity that predate the pandemic but also to reach the "new poor" affected by pandemic-induced economic disruptions.

In addition to policies aimed at sustaining incomes and food access (that is, the demand-side interventions discussed above), government policies and interventions must keep food value chains operating (that is, the supply side). As yet, food value chain operations have not suffered significant disruptions, although there were localized problems in some countries leading to temporary price increases and availability concerns. Also, hurricanes in Central America and droughts in South America in 2020, combined with the lagged effect of the pandemic, may generate some supply problems in 2021. All such developments will need to be monitored. More generally, governments should aim to make food value chains more resilient to current and future shocks. Continued support will be needed for small and medium enterprises in food systems and for small farmers through both fiscal and monetary instruments, such as tax relief and soft loans.[8]

Most LAC countries will also need to invest more in agricultural R&D, not only to improve productivity and to adapt to and mitigate climate change, but also to make food value chains more resilient to health shocks. The current pandemic, like its precursors such as SARS and avian flu, have shown the complex interactions within food chains between animal health and human health.

Management of the additional debt and monetary expansion entailed by the pandemic response will also be a challenge for several countries in the region. Those with dollarized economies will confront even tighter constraints to financing an adequate long-term response in the critical areas of health interventions, expansion of safety nets, and production and employment support.

The long-term impact of the pandemic in LAC has important implications not only for the people of the region, but also for the Sustainable Development Goals and for the planet. LAC plays a central role in terms of global food supply and environmental sustainability.[9] How the region manages the pandemic in 2020 and moves toward a sustainable transformation of the region's food systems will have global implications.

LATIN AMERICA AND THE CARIBBEAN COVID-19 TIMELINE, FEBRUARY 2020–JANUARY 2021

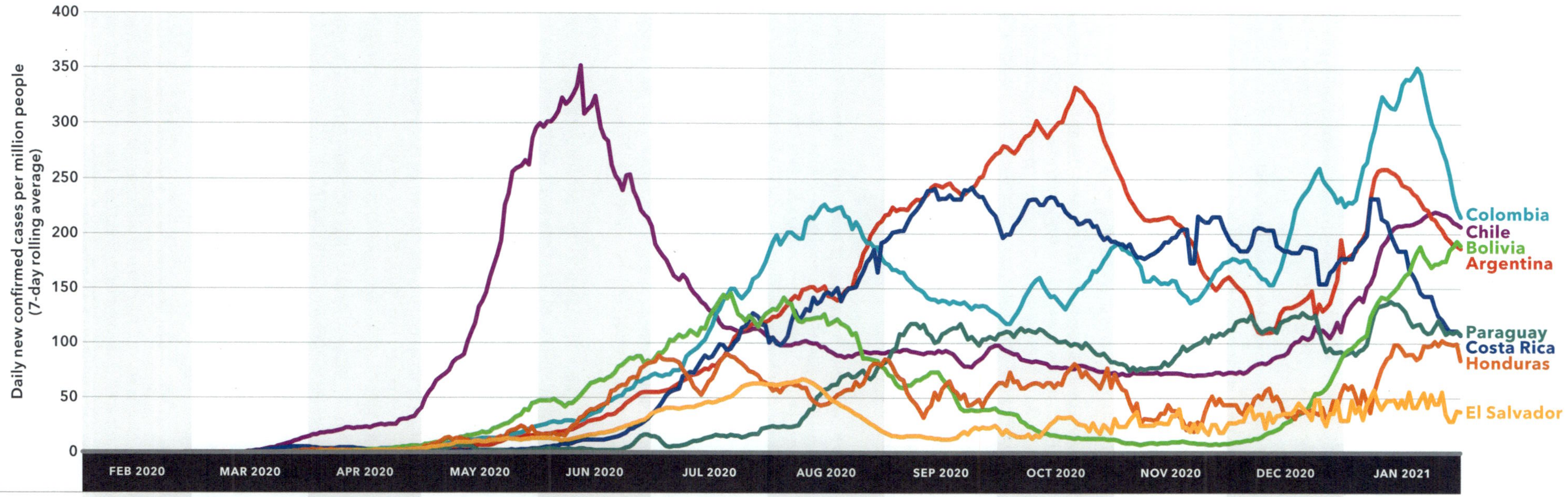

LOCKDOWNS AND HEALTH MEASURES

- **Honduras** declares state of emergency
- **Argentina**, **Paraguay**, **El Salvador** close borders, impose lockdowns
- **Bolivia** declares state of emergency, closes schools
- **Argentina**, **Bolivia** begin reopening
- **Costa Rica** restricts foreign cargo transit
- **Honduras**, **Paraguay** begin reopening
- **Costa Rica** begins reopening
- **El Salvador** reopens international airport
- **Argentina** enters new lockdown
- **Costa Rica** reopens for tourists

STIMULUS

- **Honduras** receives IMF loan
- **Bolivia** allows suspension of loan payments
- **El Salvador** offers low-interest loans for SMEs, delay on tax payments

SOCIAL PROTECTION

- **Costa Rica** launches financial support for unemployed
- **El Salvador** announces payment to about 75% of households
- **Honduras** announces fiscal program including safety nets
- **El Salvador** offers basic foodstuffs to 1.7 million families
- **Bolivia** approves payments to vulnerable groups

FOOD SECTOR INTERVENTIONS

- **Argentina** fixes maximum prices for consumer goods at March levels
- **El Salvador** removes import tariffs on staples, increases size of agricultural seed packages
- **Honduras** restricts export of red beans through July
- **Paraguay** strengthens technical assistance for small producers
- **Colombia** reduces tariffs to support agricultural sector
- **Colombia**, **El Salvador** impose price controls on basic food products

Source: For COVID-19 data, Johns Hopkins University, CSSE Database (updated March 8, 2021). For policy data, IFPRI, COVID-19 Policy Response Portal (accessed March 2021) and news reports.
Note: Countries and policies included here were selected to provide an indication of the types of actions taken in the region.

Notes

CHAPTER 1

1 FAO (Food and Agriculture Organization of the United Nations), IFAD (International Fund for Agricultural Development), UNICEF, WFP (World Food Programme), and WHO (World Health Organization), *The State of Food Security and Nutrition in the World 2020: Executive Summary* (Rome: 2020); FAO, IFAD, UNICEF, WFP, and WHO, *The State of Food Security and Nutrition in the World 2019: Safeguarding Against Slowdowns and Downturns* (Rome: 2019).

2 FAO, IFAD, UNICEF, WFP, and WHO, *The State of Food Security and Nutrition in the World 2020*.

3 Ibid.

4 Ibid.

5 IPCC (Intergovernmental Panel on Climate Change), *Climate Change and Land: Summary for Policymakers* (2020), 8.

6 S. Fan et al., "Food Systems for Human and Planetary Health: Economic Perspectives and Challenges," *Food System Economics* (forthcoming).

7 A. Dobson et al., "Ecology and Economics for Pandemic Prevention," *Science* 369, no. 6502 (2020): 379-381.

8 HLPE (High Level Panel of Experts on Food Security and Nutrition), *Food Security and Nutrition: Building a Global Narrative towards 2030* (Rome: 2020).

9 J. Swinnen and J. McDermott, eds., *COVID-19 & Global Food Security* (Washington, DC: IFPRI, 2020).

10 F. Wourterse, S. Murphy, and J. Porciello, "Social Protection to Combat Hunger," *Nature Food* 1 (2020): 517-518.

11 D. Laborde Debucquet, W. Martin, and R. Vos, "Impacts of COVID-19 on Global Poverty, Food Security and Diets," IFPRI Discussion Paper 1993 (IFPRI, Washington, DC, 2020).

12 R. Dai et al., "The Impact of COVID-19 on Small and Medium-Sized Enterprises: Evidence from Two-Wave Phone Surveys in China," CGD Working Paper (Center for Global Development, Washington, DC, 2020).

13 S. Kalantaryan and S. McMahon, *Covid-19 and Remittances in Africa*, JRC Technical Report (Luxembourg: European Union, 2020).

14 D. Laborde, W. Martin, and R. Vos, "Impacts of COVID-19 on Global Poverty, Food Security and Diets: General Equilibrium Model Estimates," *Agricultural Economics,* Special Issue on COVID-19, Global Food Systems, and Welfare, ed. J. Swinnen and R. Vos (forthcoming).

15 H. Takeshima, P. P. Zone, and I. Masias, *Monitoring the Impact of COVID-19 in Myanmar: Agricultural Equipment Retailers - November 2020 Survey Round*, MSSP Policy Note (Washington, DC, and Yangon: IFPRI, 2020).

16 K. Van Hoyweghen et al., "The Resilience of Horticultural Supply Chains to the Covid-19 Pandemic: Insights from Senegal, " *Agricultural Economics* (forthcoming, 2021).

17 A. Tesfaye and E. Habte, *How Did the Agricultural Input Supply, Dairy Value Chain and Labor Market Perform under COVID-19 Incidence in Ethiopia?*, EIAR Research Report 128 (Addis Ababa: Ethiopian Institute of Agricultural Research, 2020).

18 J. Glauber et al., "COVID-19: Trade Restrictions Are Worst Possible Response to Safeguard Food Security," *IFPRI COVID-19 Blog*, IFPRI, March 27, 2020.

19 IFPRI, COVID-19 Food Trade Policy Tracker, accessed January 18, 2021; J. Hepburn et al., *How Food Export Restrictions Could Worsen a Looming Food Crisis*, IISD/IFPRI Policy Brief (Winnipeg, Canada: International Institute for Sustainable Development, 2020).

20 Hepburn et al., *How Food Export Restrictions Could Worsen a Looming Food Crisis*.

21 FAO, *Second Rapid Assessment of Food and Nutrition Security in the Context of COVID-19 in Bangladesh: May-July 2020* (Dhaka: 2020).

22 S. Saini and S. Gupta, "Just a Call Away? Learnings from Phone Surveys in India and Nepal," *IFPRI South Asia Blog*, November 17, 2020.

23 S. Rozelle et al., "Lockdowns Are Protecting China's Rural Families from COVID-19, but the Economic Burden Is Heavy," in *COVID-19 & Global Food Security*, ed. J. Swinnen and J. McDermott (Washington, DC: IFPRI, 2020); F. Ceballos, M.A. Hernández, and C. Paz, "Short-Term Impacts of COVID-19 on Food Security and Nutrition in Rural Guatemala: Survey Evidence, *Agricultural Economics,* Special Issue on COVID-19, Global Food Systems, and Welfare, ed. J. Swinnen and R. Vos (forthcoming).

24 M. Alvi and M. Gupta, "Learning in Times of Lockdown: How Covid-19 Is Affecting Education and Food Security in India," *Food Security* 12 (2020): 793-796.

25 D.D. Headey et al., "Impacts of COVID-19 on Childhood Malnutrition and Nutrition-Related Mortality," *Lancet* 396, no. 10250 (2020): 519-521; S. Osendarp et al., "The Potential Impacts of the COVID-19 Crisis on Maternal and Child Undernutrition in Low and Middle Income Countries," *Nature Food* (under review).

26 J. Swinnen and J. McDermott, "Covid-19 and Global Food Security," *EuroChoices* 19, no. 3 (2020): 26-33.

27 K. Hirvonen, G.T. Abate, and A. de Brauw, "Food and Nutrition Security in Addis Ababa, Ethiopia during COVID-19 Pandemic: May 2020 Report," ESSP Working Paper (IFPRI, Washington, DC, 2020).

28 J. Swinnen and R. Vos, "COVID-19, Global Food Systems and Welfare: Key Insights," *Agricultural Economics,* Special Issue on COVID-19, Global Food Systems, and Welfare, ed. J. Swinnen and R. Vos (forthcoming).

29 IFPRI, COVID-19: Measuring Impacts and Prioritizing Policies for Recovery (webpage), accessed Feb. 2021.

30 A. Madgavkar et al., "COVID-19 and Gender Equality: Countering the Regressive Effects," McKinsey Global Institute, July 15, 2020.

31 K. Limaye, D. Chandurkar, and N. Choudhury, "Lockdown and Rural Distress: Highlights from Phone Surveys of 5,000 Households in 12 Indian States," *IFPRI South Asia Blog*, July 13, 2020.

32 J.M. Aguero, "COVID-19 and the Rise of Intimate Partner Violence," *World Development* 137 (January 2021): 105217.

33 C. Ragasa and I. Lambrecht, "COVID-19 and the Food System: Setback or Opportunity for Gender Equality?," *Food Security* 12 (2020): 877-880.

34 GLOPAN (Global Panel on Agriculture and Food Systems for Nutrition), *Future Food Systems: For People, Our Planet, and Prosperity* (London: 2020).

35 IFPRI, COVID-19 Policy Response Portal, accessed March 12, 2021.

36 K.A. Abay et al., "COVID-19 and Food Security in Ethiopia: Do Social Protection Programs Protect?," World Bank Policy Research Working Paper 9475 (World Bank, Washington, DC, 2020); F. Ceballos, S. Kannan,

and B. Kramer, "Impacts of a National Lockdown on Smallholder Farmers' Income and Food Security: Empirical Evidence from Two States in India," *World Development* 136 (December 2020): 105069.

37 A. Mukherjee, "Digital Cash Transfers for Stranded Migrants: Lessons from Bihar's COVID-19 Assistance Program," *CGD Notes* (Washington, DC: Center for Global Development, 2020).

38 A. Kennedy and D. Resnick, *From Farm to Table: How Are Governments Keeping Food Systems Functioning during COVID-19?*, COVID-19 Policy Response Portal Project Note 1 (Washington, DC: IFPRI, 2020).

39 Swinnen and Vos, "COVID-19, Global Food Systems and Welfare: Key Insights."

40 K. Van Hoyweghen et al., "Resilience of Global and Local Value Chains to the COVID-19 Pandemic: Survey Evidence from Vegetable Value Chains in Senegal," *Agricultural Economics*, Special Issue on COVID-19, Global Food Systems, and Welfare, ed. J. Swinnen and R. Vos (forthcoming).

41 L.A. Post et al., "Dynamic Metrics for Public Health Surveillance Are Imperative to Gain Control of the COVID-19 Pandemic in America: Longitudinal Trend Analysis," *Journal of Medical Internet Research* 22, no. 12 (2020): e24286.

42 IFPRI, Food Security Portal, accessed March 12, 2021.

43 M.T. Ruel and H. Alderman, "Nutrition-Sensitive Interventions and Programmes: How Can They Help to Accelerate Progress in Improving Maternal and Child Nutrition?," *Lancet* 382, no. 9891: 536–551.

FINANCING THE TRANSFORMATION TO HEALTHY, SUSTAINABLE, AND EQUITABLE FOOD SYSTEMS

1 See for example, FOLU (Food and Land Use Coalition), *Growing Better: Ten Critical Transitions to Transform Food and Land Use*, Global Consultation Report of the Food and Land Use Coalition (2019); J. von Braun et al., *Ending Hunger by 2030: Policy Actions and Costs*, Policy Brief (Bonn: Center for Development Research, 2020); E. Díaz-Bonilla and S. Saravia-Matus, "Los objetivos de desarrollo sostenible y el desarrollo rural en América Latina y el Caribe: Reflexiones sobre costos y financiamientos," *2030 Alimentación, agricultura y desarrollo rural en América Latina y el Caribe* 33 (Santiago, Chile: FAO, 2019).

2 These core objectives are being promoted by the UN Food System Summit's Finance Lever. See UNFSS Finance Lever Secretariat, "Transforming the Financial System for Healthy People, Healthy Economy, Healthy Planet" (unpublished paper, 2021).

3 OECD (Organisation for Economic Co-operation and Development), *Agricultural Policy Monitoring and Evaluation* (Paris: 2020); D. Laborde, A. Mamun, W. Martin, V. Piñeiro, and R. Vos, "Agricultural Subsidies and Global Greenhouse Emissions," *Nature Communications* (forthcoming, 2021)

4 See for example, D. Laborde, W. Martin, and R. Vos, "Impacts of COVID-19 on Global Poverty, Food Security, and Diets," *Agricultural Economics* (forthcoming).

5 E. Díaz-Bonilla, "Fiscal and Monetary Responses to the COVID-19 Pandemic: Some Thoughts for Developing Countries and the International Community," in *COVID-19 & Global Food Security*, ed. J. Swinnen and J. McDermott (Washington, DC: IFPRI, 2020).

6 Data are from OECD's reporting system for development assistance, "Official Development Assistance," accessed Feb 28, 2021, oecd-ilibrary.org. For further analysis, see I. Bharali et al., *The Financing Landscape for Agricultural Development: An Assessment of External Financing Flows to Low- and Middle-Income Countries and of the Global Aid Architecture* (Durham, NC: Duke Sanford World Food Policy Center, 2020).

7 E. Díaz-Bonilla, "The Future Ain't What It Used to Be: Growth Models, Structural Change, and History," MTID Working Paper (IFPRI, Washington, DC, 2016).

8 See, for example, E. Díaz-Bonilla et al., *Financing "A Sustainable Food Future,"* T20 Argentina: Food Security and Sustainable Development Task Force Brief (Buenos Aires: Consejo Argentino para las Relaciones Internacionales and CIPPEC, 2018).

CHAPTER 2

1 L. Moorty et al., *A Policy Framework for Mitigating the Impact of the COVID-19 Crisis*, Policy Working Paper (World Bank, Washington, DC, 2020).

2 Z. Tufecki, "This Overlooked Variable Is the Key to the Pandemic," *Atlantic*, September 30, 2020.

3 Oxford Supertracker, Global Directory for COVID Policy Trackers and Surveys, "Policy Trackers," https://supertracker.spi.ox.ac.uk/policy-trackers/.

4 S. Allin et al., *Comparing Policy Responses to COVID-19 among Countries in the Latin American and Caribbean (LAC) Region* (Washington, DC: World Bank, 2020).

5 K.A. Abay et al., "Ethiopia's Social Safety Net Effective in Limiting COVID-19 Impacts on Rural Food Insecurity," *IFPRI COVID-19 Blog*, IFPRI, January 28, 2021.

6 J. McDermott and A. de Brauw, "National Food Systems: Inclusive Transformation for Healthier Diets," in *2020 Global Food Policy Report: Building Inclusive Food Systems* (Washington, DC: IFPRI, 2020).

7 D. Resnick, "COVID-19 Lockdowns Threaten Africa's Vital Informal Urban Food Trade," *IFPRI COVID-19 Blog*, March 31, 2020.

8 K.S. Andam et al., "Estimating the Economic Costs of COVID-19 in Nigeria," NSSP Working Paper (IFPRI, Abuja, 2020).

9 J. Harris, "Diets in a Time of Coronavirus: Don't Let Vegetables Fall Off the Plate," *IFPRI COVID-19 Blog*, April 13, 2020; S. Tamru, K. Hirvonen, and B. Minten, "Impacts of the COVID-19 Crisis on Vegetable Value Chains in Ethiopia," *IFPRI COVID-19 Blog*, April 13, 2020.

10 T. Reardon and J. Swinnen, "COVID-19 and Resilience Innovations in Food Supply Chains," *IFPRI COVID-19 Blog*, July 6, 2020.

11 OECD (Organisation for Economic Co-operation and Development), "The Territorial Impact of COVID-19: Managing the Crisis across Levels of Government," Tackling Coronavirus (COVID-19): Contributing to a Global Effort (webpage), Nov. 2020 update.

12 A. Vassall et al., "Integrating Economic and Health Evidence to Inform Covid-19 Policy in Low- and Middle-Income Countries," [version 1; peer review: 1 approved with reservations] *Wellcome Open Research* 5 (2020): 272.

13 S. Torres-Rueda et al., "The Health Sector Cost of Different Policy Responses to COVID-19 in Low- and Middle-Income Countries," medRxiv, August 25, 2020.

14 Ibid.

15 World Bank, *Poverty and Shared Prosperity 2020: Reversals of Fortune* (Washington, DC: 2020).

16 World Bank, *Global Economic Prospects, January 2021* (Washington, DC: 2021).

17 Ibid.

18 Ibid.

19 World Bank, *Poverty and Shared Prosperity 2020.*

20 S. Sweeney et al., "Exploring Equity in Health and Poverty Impacts of Control Measures for SARS-CoV-2 in Six Countries," *Lancet* (preprint, Feb. 2021).

21 L. Moorty et al., *A Policy Framework for Mitigating the Impact of the COVID-19 Crisis*.

22 D. Laborde and C. Smaller, "What Would It Cost to Avert the COVID-19 Hunger Crisis?," Ceres2030 Report, 2020.

23 Vassall et al., "Integrating Economic and Health Evidence to Inform Covid-19 Policy in Low- and Middle-Income Countries."

24 A. Kennedy and D. Resnick, *From Farm to Table: How Are Governments Keeping Food Systems Functioning during COVID-19?*, COVID-19 Policy Response Portal Project Note No. 1 (Washington, DC: IFPRI, 2020).

25 S. Knott, "In Ghana, Drones Help Combat COVID-19," Voice of America News, April 29, 2020.

26 ILO (International Labour Organization), *Extending Social Protection to Informal Workers in the COVID-19 Crisis: Country Responses and Policy Considerations*, ILO Brief (Geneva: 2020).

27 WEF (World Economic Forum), *Global Technology Governance Report 2021* (Geneva: 2020).

28 L. Alfers, G. Ismail, and M. Valdivia, *Informal Workers and the Social Protection Response to COVID-19: Who Got Relief? How? And Did It Make a Difference?* (Manchester, UK: Women in Informal Employment: Globalizing and Organizing [WIEGO], 2020).

29 IFPRI, COVID-19 Policy Response Portal, accessed Feb. 2021.

30 OECD, "The Territorial Impact of COVID-19: Managing the Crisis across Levels of Government."

31 UNCDF (Unlocking Public and Private Finance for the Poor), "Ugandan Local Governments, the Driving Force Behind the Covid-19 Response," April 9, 2020.

32 Kampala Capital City Authority, "Online Market," March 28, 2020.

33 E. Onyeji and N. Adebowale, "Nigerian Govt Warns against COVID-19 'Politics' in Cross Rivers, Kogi," *Premium Times*, July 7, 2020.

34 E. Londoño, M. Andreoni, and L. Casado, "Bolsonaro, Isolated and Defiant, Dismisses Coronavirus Threat to Brazil," *New York Times*, June 18, 2020; E. Londoño, M. Andreoni, and L. Casado, " 'Playing with Lives': Brazil's Covid Vaccine Plan Is Mired in Chaos," *New York Times*, December 14, 2020.

35 J. Gettleman et al., "The Virus Trains: How Lockdown Chaos Spread COVID-19 across India," *New York Times*, December 15, 2020.

36 See examples of social protection (India) and food aid (Bangladesh) in P. Fang, A. Kennedy, and D. Resnick, *Scaling Up and Sustaining Social Protection under COVID-19*, COVID-19 Policy Response Portal Project Note No. 3 (Washington, DC: IFPRI, 2020).

37 A. de Waal and P. Richards, "Coronavirus: Why Lockdowns May Not Be the Answer in Africa," BBC News, April 14, 2020.

38 HRW (Human Rights Watch), "Covid-19 Triggers Wave of Free Speech Abuse: Scores of Countries Target Media, Activists, Medics, Political Opponents," HRW News, February 11, 2021.

39 G. Capano et al., "Mobilizing Policy (In)Capacity to Fight COVID-19: Understanding Variations in State Responses," *Policy and Society* 39, no. 3 (2020): 285–308.

40 T.O. Ali, M. Hassan, and N. Hossain, "The Moral and Political Economy of the Pandemic in Bangladesh: Weak States and Strong Societies during COVID-19," *World Development* 137 (2021): 105216.

41 D. Resnick, "Trust in Science and in Government Plays a Crucial Role in COVID-19 Response," *IFPRI COVID-19 Blog*, IFPRI, June 10, 2020.

42 B. Ojikutu, K. Stephenson, K. Mayer, and K. Emmons, "Building Trust in COVID-19 Vaccines and Beyond through Authentic Community Investment," *American Journal of Public Health* 111, no. 3 (2021): 366–368.

43 IRC (International Rescue Committee), *The Road to COVID-19 Immunity: Building Trust and Combating Misinformation* (New York: 2021).

CHAPTER 3

1 FAO (Food and Agriculture Organization of the United Nations), IFAD (International Fund for Agricultural Development), UNICEF, WFP (World Food Programme), and WHO (World Health Organization), *The State of Food Security and Nutrition in the World 2020: Transforming Food Systems for Affordable Healthy Diets* (Rome: FAO, 2020).

2 D. Laborde et al., "COVID-19 Pandemic Leads to Greater Depth of Unaffordability of Healthy and Nutrient Adequate Diets in Low- and Middle-Income Countries," *Nature Food* (under review).

3 I. Goldin and R. Muggah, "COVID-19 Is Increasing Multiple Kinds of Inequality: "Here's What We Can Do about It," World Economic Forum, October 9, 2020.

4 D. Headey and M. Ruel, "The COVID-19 Nutrition Crisis: What to Expect and How to Protect," in *COVID-19 & Global Food Security*, ed. J. Swinnen and J. McDermott (Washington, DC: IFPRI, 2020).

5 S. Osendarp et al., "The Potential Impacts of the COVID-19 Crisis on Maternal and Child Undernutrition in Low and Middle Income Countries," *Nature Food* (under review).

6 B. Popkin, C. Corvalan, and L.M. Grummer-Strawn, "Dynamics of the Double Burden of Malnutrition and the Changing Nutrition Reality," *Lancet* 395, no. 10217 (2020): 65–74.

7 P.C. Calder, "Nutrition, Immunity and COVID-19," *BMJ Nutrition, Prevention & Health* 3 (2020): 74–92.

8 Global Burden of Diseases (GBD) Diet Collaborators 2017, "Health Effects of Dietary Risks in 195 Countries, 1990–2017: A Systematic Analysis for the Global Burden of Disease Study 2017," *Lancet* 393, no. 10184 (2019): 1958–1972.

9 Development Initiatives, "Action on Equity to End Malnutrition," in *2020 Global Nutrition Report* (Bristol, UK: 2020).

10 K.D. Hall et al., "Ultra-processed Diets Cause Excess Calorie Intake and Weight Gain: An Inpatient Randomized Controlled Trial of Ad Libitum Food Intake," *Cell Metabolism* 30, no. 1 (2019): 67–77; M. Bonaccio et al., "Ultra-processed Food Consumption Is Associated with Increased Risk of All-Cause and Cardiovascular Mortality in the Moli-sani Study," *American Journal of Clinical Nutrition* 113, no. 2 (2020): 446–455.

11 GLOPAN (Global Panel on Agriculture and Food Systems for Nutrition), *Future Food Systems: For People, Our Planet, and Prosperity* (London: 2020).

12 J. Swinnen and J. McDermott, eds., *COVID-19 & Global Food Security* (Washington, DC: IFPRI, 2020).

13 UN Women, *From Insights to Action: Gender Equality in the Wake of Covid-19* (New York: 2020).

14 A. Tesfaye, Y. Habte, and B. Minten, "COVID-19 Is Shifting Consumption and Disrupting Dairy Value Chains in Ethiopia," in *COVID-19 & Global Food Security*; K. Hirvonen, G. Abate, and A. de Brauw, "Survey Suggests Rising Risk of Food and Nutrition Insecurity in Addis Ababa, Ethiopia, as COVID-19 Restrictions Continue," in *COVID-19 & Global Food Security*.

15 C. Hawkes and C. Gallagher Squires, "Building Back Better from COVID-19 for Obesity and Undernutrition: Time to Implement a 'Double Duty' Stimulus Package for Healthier Food Systems," *Nature Food* (under review).

16 D. Laborde, W. Martin, and R. Vos, "Estimating the Poverty Impact of COVID-19: The MIRAGRODEP and POVANA Frameworks," IFPRI Technical Note (Washington, DC: IFPRI, 2020).

17 Goldin and Muggah, "COVID-19 Is Increasing Multiple Kinds of Inequality."

18 A. Josephson, T. Kilic, and J. Michler, "Socioeconomic Impacts of COVID-19 in Four African Countries," Policy Research Working Paper 9466 (World Bank, Washington, DC, 2020); D. Headey et al., "Waves of Disease, Waves of Poverty: New Evidence on the Economic Impacts of COVID-19 in Myanmar," *IFPRI COVID-19 Blog*, December 1, 2020; H. Alderman et al., "Short-Term Evidence on Wellbeing of Rural Ethiopian Households during the COVID-19 Pandemic," SPIR Learning Brief-4 (Washington, DC: IFPRI, 2020); M.K. Kansiime et al., "COVID-19 Implications on Household Income and Food Security in Kenya and Uganda: Findings from a Rapid Assessment," *World Development* 137 (2021): 105199; M. Amare et al., "Impacts of COVID-19 on Food Security: Panel Data Evidence from Nigeria," IFPRI Discussion Paper 01956 (IFPRI, Washington, DC, 2020); A. Ahmed, M. Bakhtiar, and J. Ghostlaw, "Assessing COVID-19's Effects on Socioeconomic and Food Security Characteristics through Phone Surveys in Bangladesh" (Dhaka: IFPRI, 2020).

19 Hirvonen et al., "Survey Suggests Rising Risk of Food and Nutrition Insecurity in Addis Ababa, Ethiopia, as COVID-19 Restrictions Continue"; Headey et al., "Waves of Disease, Waves of Poverty: New Evidence on the Economic Impacts of COVID-19 in Myanmar"; Alderman et al., "Short-Term Evidence on Wellbeing of Rural Ethiopian Households during the COVID-19 Pandemic; Kansiime et al., "COVID-19 Implications on Household Income and Food Security in Kenya and Uganda"; Ahmed et al., "Assessing COVID-19's Effects on Socioeconomic and Food Security Characteristics through Phone Surveys in Bangladesh"; P. Menon et al., "COVID-19 and Nutrition in India: The Challenges of Preserving Progress through a Period of Crisis," *Nature Food* (under review); C. Chege et al., *Impact of COVID-19 on Diets of Poor Consumers in Africa: Evidence from the Slums of Nairobi, Kenya* (Cali, Colombia: International Center for Tropical Agriculture [CIAT], 2020).

20 Ahmed et al, "Assessing COVID-19's Effects on Socioeconomic and Food Security Characteristics through Phone Surveys in Bangladesh."

21 Headey and Ruel, "The COVID-19 Nutrition Crisis: What to Expect and How to Protect."

22 Hawkes and Gallagher Squires, "Building Back Better from COVID-19 for Obesity and Undernutrition."

23 A. de Brauw, K. Hirvonen, and G. Abate, "Survey: Despite COVID-19, Food Consumption Remains Steady in Addis Ababa, Ethiopia," *IFPRI COVID-19 Blog*, Oct. 2, 2020; M. Bakhtiar, A. Ahmed, D. Gilligan, J. Hodddinott, and S. Roy, "Assessing Impacts of COVID-19 in Bangladesh: Preliminary Results from Recent Phone Surveys," PowerPoint presentation, IFPRI, 2021.

24 U. Gentilini, M. Almenfi, and P. Dale, "Social Protection and Jobs Responses to COVID-19: A Real-Time Review of Country Measures," living paper version 14, December 2020 (Washington, DC: World Bank, 2020).

25 S. Abed, "Opinion: The Long Shadow of COVID-19 on Extreme Poverty," Devex, October 27, 2020; N. Kumar et al., "Toward Inclusive Food Systems: Pandemics, Vulnerable Groups, and the Role of Social Protection " in this volume.

26 I.D. Brouwer, J. McDermott, and R. Ruben, "Food Systems Everywhere: Improving Relevance in Practice," *Global Food Security* 26 (2020): 100398; HLPE (High Level Panel of Experts), *Nutrition and Food Systems: A Report by the High Level Panel of Experts on Food Security and Nutrition of the Committee on World Food Security* (Rome: 2017); FAO, IFAD, UNICEF, WFP, and WHO, *State of Food Security and Nutrition in the World 2020*; GLOPAN, *Future Food Systems*.

27 C. Hawkes, M.T. Ruel, L. Salm, B. Sinclair, and F. Branco, "Double-Duty Actions: Seizing Programme and Policy Opportunities to Address Malnutrition in All Its Forms," *Lancet* 395, no. 10218 (2020): 142–155.

28 Gentilini, Almenfi, and Dale, "Social Protection and Jobs Responses to COVID-19."

29 Hawkes and Gallagher Squires, "Building Back Better from COVID-19 for Obesity and Undernutrition"; M. Hidrobo et al., "Gender-Sensitive Social Protection: A Critical Component of the COVID-19 Response in Low- and Middle-Income Countries," IFPRI Issue Brief (Washington, DC: IFPRI, 2020); D. Gilligan, "Social Safety Nets Are Crucial to the COVID-19 Response. Some Lessons to Boost Their Effectiveness," *IFPRI COVID-19 Blog*, June 18, 2020.

30 M. Springmann et al., "Health and Nutritional Aspects of Sustainable Diet Strategies and Their Association with Environmental Impacts: A Global Modelling Analysis with Country-Level Detail," *Lancet Planetary Health* 2, no. 10 (2018): e451–461; A. Herforth, M. Arimond, C. Álvarez-Sánchez, J. Coates, K. Christianson, and E. Muehlhoff, "A Global Review of Food-Based Dietary Guidelines," *Advances in Nutrition* 10, no. 4 (July 2019): 590–605.

31 Herforth et al., "A Global Review of Food-Based Dietary Guidelines."

32 B.M. Popkin, C. Corvalan, and L.M. Grummer-Strawn, "Dynamics of the Double Burden of Malnutrition and the Changing Nutrition Reality," *Lancet* 395, no. 10217 (2020): 65–74.

33 A. Herforth and S. Ahmed, "The Food Environment, Its Effects on Dietary Consumption, and Potential for Measurement within Agriculture-Nutrition Interventions," *Food Security* 7 (2015): 505–520.

34 J. Mandle et al., "Nutrition Labelling: A Review of Research on Consumer and Industry Response in the Global South," *Global Health Action* 8 (2015): 25912.

35 WHO, *Marketing of Food and Non-alcoholic Beverages to Children: Report of a WHO Forum and Technical Meeting*, Oslo, May 2-5, 2006 (Geneva: 2006).

36 S. Anzman-Frasca et al., "Changes in Children's Meal Orders Following Healthy Menu Modifications at a Regional US Restaurant Chain," *Obesity* 23 (2015): 1055–1062.

37 J.D. Jensen and S. Smed, "State-of-the-Art for Food Taxes to Promote Public Health," *Proceedings of the Nutrition Society* 77, no. 2 (2018): 100–105.

38 L.E.W. Olsho, J.A. Klerman, P.E. Wilde, and S. Bartlett, "Financial Incentives Increase Fruit and Vegetable Intake among Supplemental Nutrition Assistance Program Participants: A Randomized Controlled Trial of the USDA Healthy Incentives Pilot," *American Journal of Clinical Nutrition* 104, no. 2 (2016): 423–435.

39 C. Young et al., "Supporting Engagement, Adherence, and Behavior Change in Online Dietary Interventions," *Journal of Nutrition Education and Behavior* 51, no. 6 (2019): 719–739.

40 A.M. Müller et al., "The Effectiveness of e- & mHealth Interventions to Promote Physical Activity and Healthy Diets in Developing Countries: A Systematic Review," *International Journal of Behavorial Nutrition and Physical Activity* 13, no. 109 (2016).

41 E.P. Abril and P. R. Dempsey, "Outcomes of Healthy Eating Ad Campaigns: A Systematic Review," *Progress in Cardiovascular Diseases* 62, no. 1 (2019): 39–43.

42 E. Robinson et al., "What Everyone Else Is Eating: A Systematic Review and Meta-Analysis of the Effect of Informational Eating Norms on Eating Behavior," *Journal of the Academy of Nutrition and Dietetics* 114, no. 3 (2014): 414–429.

43 A. Webb Girard et al., "A Scoping Review of Social-Behaviour Change Techniques Applied in Complementary Feeding Interventions," *Maternal and Child Nutrition* 16, no. 1 (2020): e12882.

44 J. Leroy and M. Ruel, "Covid-19 and the Irreversible Marks Undernutrition Will Leave in Its Wake," *Outlook POSHAN*, December 10, 2020.

BOX 2

a A. Josephson, T. Kilic, and J. Michler, "Socioeconomic Impacts of COVID-19 in Four African Countries," Policy Research Working Paper 9466 (World Bank, Washington, DC, 2020); D. Headey et al., "Waves of Disease, Waves of Poverty: New Evidence on the Economic Impacts of COVID-19 in Myanmar," *IFPRI COVID-19 Blog*, December 1, 2020; A. de Brauw, K. Hirvonen, and G. Abate, "Survey: Despite COVID-19, Food Consumption Remains Steady in Addis Ababa, Ethiopia," *IFPRI COVID-19 Blog*, October 2, 2020.

b H. Rahman and I. Matin, "Livelihoods, Coping, and Support during Covid-19 Crisis" (Presentation, Dhaka, Bangladesh, 2020).

c A. Ahmed, M. Bakhtiar, and J. Ghostlaw, "Assessing COVID-19's Effects on Socioeconomic and Food Security Characteristics through Phone Surveys in Bangladesh" (Dhaka: IFPRI, 2020).

d K. Hirvonen, G. Abate, and A. de Brauw, "Survey Suggests Rising Risk of Food and Nutrition Insecurity in Addis Ababa, Ethiopia, as COVID-19 Restrictions Continue," in *COVID-19 & Global Food Security*, ed. J. Swinnen and J. McDermott (Washington, DC: IFPRI, 2020).

e Josephson et al., "Socioeconomic Impacts of COVID-19 in Four African Countries."

f Headey et al., "Waves of Disease, Waves of Poverty: New Evidence on the Economic Impacts of COVID-19 in Myanmar."

g Josephson et al., "Socioeconomic Impacts of COVID-19 in Four African Countries."

h G. Pastori et al., "Diet Quality in the Context of COVID-19 Pandemic–Nigeria" (Wageningen, Netherlands: WUR, 2020).

i M.K. Kansiime et al., "COVID-19 Implications on Household Income and Food Security in Kenya and Uganda" *World Development* 137 (January 2021).

j Hirvonen et al., "Survey Suggests Rising Risk of Food and Nutrition Insecurity in Addis Ababa, Ethiopia, as COVID-19 Restrictions Continue"; Headey et al., "Waves of Disease, Waves of Poverty: New Evidence on the Economic Impacts of COVID-19 in Myanmar"; J.D. Hamadani, M.I. Hasan, A.J. Baldi, et al., "Immediate Impact of Stay-at-Home Orders to Control COVID-19 Transmission on Socioeconomic Conditions, Food Insecurity, Mental Health, and Intimate Partner Violence in Bangladeshi Women and Their Families: An Interrupted Time Series," *Lancet Global Health* 8, no. 11 (2020): e1380-89.

k Hirvonen et al., "Survey Suggests Rising Risk of Food and Nutrition Insecurity in Addis Ababa, Ethiopia, as COVID-19 Restrictions Continue"; Headey et al., "Waves of Disease, Waves of Poverty: New Evidence on the Economic Impacts of COVID-19 in Myanmar"; H. Alderman et al., "Short-Term Evidence on Wellbeing of Rural Ethiopian Households during the COVID-19 Pandemic," SPIR Learning Brief-4 (Washington, DC: IFPRI, 2020); M.K. Kansiime et al., "COVID-19 Implications on Household Income and Food Security in Kenya and Uganda"; Ahmed et al., "Assessing COVID-19's Effects on Socioeconomic and Food Security Characteristics through Phone Surveys in Bangladesh"; P. Menon et al., "COVID-19 and Nutrition in India: The Challenges of Preserving Progress through a Period of Crisis," *Nature Food* (under review); C. Chege et al., *Impact of COVID-19 on Diets of Poor Consumers in Africa: Evidence from the Slums of Nairobi, Kenya* (Cali, Colombia: International Center for Tropical Agriculture [CIAT], 2020).

l Headey et al., "Waves of Disease, Waves of Poverty: New Evidence on the Economic Impacts of COVID-19 in Myanmar."

m Pastori et al., "Diet Quality in the Context of COVID-19 Pandemic–Nigeria."

n Pastori et al., "Diet Quality in the Context of COVID-19 Pandemic–Nigeria"; G. Pastori et al., "Diet Quality in the Context of COVID-19 Pandemic–Vietnam" (Wageningen, Netherlands: WUR, 2020).

o C. Hawkes and C. Gallagher Squires, "Building Back Better from COVID-19 for Obesity and Undernutrition: Time to Implement a 'Double Duty' Stimulus Package for Healthier Food Systems," *Nature Food* (under review).

CHAPTER 4

1 W. Zhang et al., "Ecosystem Services and Dis-services to Agriculture," *Ecological Economics* 64, no. 2 (2007): 253-260.

2 P. Lindsey et al., "Conserving Africa's Wildlife and Wildlands through the COVID-19 Crisis and Beyond," *Nature Ecolology and Evolution* 4 (Oct. 2020): 1300-1310.

3 TEEB (The Economics of Ecosystems and Biodiversity), *Measuring What Matters in Agriculture and Food Systems: A Synthesis of the Results and Recommendations of TEEB for Agriculture and Food's Scientific and Economic Foundations Report* (Geneva: UN Environment, 2018).

4 CGIAR Research Program on Water, Land and Ecosystems (WLE), *Ecosystem Services and Resilience Framework* (Colombo: International Water Management Institute, 2014).

5 P.G. Curtis et al., "Classifying Drivers of Global Forest Loss," *Science* 361, no. 6407 (2018): 1108-1111.

6 I. Small, J. van der Meer, and R.E.G. Upshur, "Acting on an Environmental Health Disaster: The Case of the Aral Sea," *Environmental Health Perspectives* 109, no. 6 (2001): 547-549; P. Micklin and N.V. Aladin, "Reclaiming the Aral Sea," *Scientific American* 298, no. 4 (2008): 64-71; H. Gao et al., "On the Causes of the Shrinking of Lake Chad," *Environmental Research Letters* 6, no. 3 (2011): 034021.

7 F. Sánchez-Bayo and K.A.G. Wyckhuys "Worldwide Decline of the Entomofauna: A Review of Its Drivers," *Biological Conservation* 232 (April 2019): 8-27.

8 F. Geiger et al., "Persistent Negative Effects of Pesticides on Biodiversity and Biological Control Potential on European Farmland," *Basic and Applied Ecology* 11, no. 2 (2010): 97-105; K.L. Bassil et al., "Cancer Health Effects of Pesticides: Systematic Review," *Canadian Family Physician* 53, no. 10 (2007): 1704-1711.

9 L. Montanarella, R. Scholes, and A. Brainich, eds., *IPBES Assessment Report on Land Degradation and Restoration* (Bonn: Secretariat of the Intergovernmental Science-Policy Platform on Biodiversity and Ecosystem Services, 2018).

10 M.K. van Ittersum et al., "Can Sub-Saharan Africa Feed Itself?," *Proceedings of the National Academy of Sciences* 113, no. 52 (2016): 14964-14969.

11 J.R. Rohr et al., "Emerging Human Infectious Diseases and the Links to Global Food Production," *Nature Sustainability* 2 (June 2019): 445-456.

12 UNEP (United Nations Environment Programme) and ILRI (International Livestock Research Institute), *Preventing the Next Pandemic: Zoonotic Diseases and How to Break the Chain of Transmission* (Nairobi: UNEP, 2020).

13 FAO (Food and Agriculture Organization of the United Nations), *The State of the World's Biodiversity for Food and Agriculture* (Rome: FAO Commission on Genetic Resources for Food and Agriculture Assessments, 2019); P. Dasgupta, "The Economics of Biodiversity: The Dasgupta Review." (London: HM Treasury, 2021).

14 P. Dasgupta, "Environment: How Best to Face the Coming Storm," *Science* 305 (2004): 1716.

15 Montanarella et al., *IPBES Assessment Report on Land Degradation and Restoration.*

16 A.B. Durning, *Poverty and the Environment: Reversing the Downward Spiral*, Worldwatch Paper 92 (Washington DC: Worldwatch Institute, 1989); J.D. Sachs, "Sustainable Development," *Science* 304, no. 5617 (2004): 649.

17 T. Tscharntke et al., "Landscape Perspectives on Agricultural Intensification and Biodiversity: Ecosystem Service Management," *Ecology Letters* 8, no. 8 (2005): 857–874.

18 W. Zhang et al., "Biodiversity and Ecosystem Services," in *Sustainable Food and Agriculture: An Integrated Approach*, ed. C. Campanhola and S. Pandey (Cambridge, MA: Academic Press, 2019).

19 B.T. van Zanten, P.H. Verburg, M.J. Koetse, and P.J. van Beukering, "Preferences for European Agrarian Landscapes: A Meta-analysis of Case Studies," *Landscape and Urban Planning* 132 (2014): 89–101.

20 R. Seppelt et al., "Deciphering the Biodiversity–Production Mutualism in the Global Food Security Debate," *Trends in Ecology & Evolution* 35, no. 11 (2020): 1011–1020.

21 A. Haines, "Health in the Bioeconomy," *Lancet Planetary Health* 5, no. 1 (2021): E4–E5; A. Parodi et al., "The Potential of Future Foods for Sustainable and Healthy Diets," *Nature Sustainability* 1, no. 12 (2018): 782–789.

22 TEEB, *Measuring What Matters in Agriculture and Food Systems*.

23 R. Seppelt et al., "Deciphering the Biodiversity–Production Mutualism in the Global Food Security Debate," *Trends in Ecology & Evolution* 35, no. 11 (2020): 1011–1020.

24 K. Andersson, *Landscape Governance: What Is It, and What Is It Good For?* PIM Synthesis Brief (Washington, DC: IFPRI, forthcoming).

25 K. Coleman, M.J. Stern, and J. Widmer, "Facilitation, Coordination, and Trust in Landscape-Level Forest Restoration," *Journal of Forestry* 116, no. 1 (2018): 41–46.

26 M.A.F. Ros-Tonen, J. Reed, and T. Sunderland, "From Synergy to Complexity: The Trend toward Integrated Value Chain and Landscape Governance," *Environmental Management* 62 (May 2018): 1–14.

27 S. Engel, S. Pagiola, and S. Wunder, "Designing Payments for Environmental Services in Theory and Practice: An Overview of the Issues," *Ecological Economics* 65, no. 4 (2008): 663–674.

28 N. Uphoff, "Grassroots Organizations and NGOs in Rural Development: Opportunities with Diminishing States and Expanding Markets," *World Development* 21, no. 4 (1993): 607–622; J. Kerr, "Watershed Management: Lessons from Common Property Theory," *International Journal of the Commons* 1, no. 1 (2007): 89–110.

29 Kerr, "Watershed Management: Lessons from Common Property Theory"; J.F. Warner, "More Sustainable Participation? Multi-Stakeholder Platforms for Integrated Catchment Management," *International Journal of Water Resources Development* 22, no. 1 (2006): 15–35; A. Akhmouch and D. Clavreul, "Stakeholder Engagement for Inclusive Water Governance: 'Practicing What We Preach' with the OECD Water Governance Initiative," *Water* (Switzerland) 8, no. 5 (2016): 1–17.

30 C. Hawkes and J. Voegele, "Our Food System Is Broken: Here Are 3 Ways to Fix It," World Economic Forum, January 19, 2018.

CHAPTER 5

1 L. Wade, "An Unequal Blow," *Science* 368, no. 6492 (2020): 700–703.

2 T. Reardon, M. F. Bellemare, and D. Zilberman, "How COVID-19 May Disrupt Food Supply Chains in Developing Countries," in *COVID-19 & Global Food Security*, ed. J. Swinnen and J. McDermott (Washington, DC: IFPRI, 2020).

3 United Nations Human Settlements Programme (UN-Habitat), "Opinion: COVID-19 Demonstrates Urgent Need for Cities to Prepare for Pandemics," June 15, 2020.

4 C.S. Brown, M. Ravallion, and D. van de Walle, "Can the World's Poor Protect Themselves from the New Coronavirus?," NBER Working Paper 27200 (National Bureau of Economic Research, Cambridge, MA, 2020).

5 N.D. Sheth, "India's Rustic Mom-and-Pop Stores Make Comeback amid Pandemic," NikkeiAsia, May 14, 2020.

6 J. Thurlow, "COVID-19 Lockdowns Have Imposed Substantial Economic Costs on Countries in Africa," in *COVID-19 & Global Food Security*.

7 World Bank, *The State of Social Safety Nets 2018* (Washington, DC: 2018).

8 B. Minten, B. Mohammed, and S. Tamru, "Emerging Medium-Scale Tenant Farming, Gig Economies, and the COVID-19 Disruption: The Case of Commercial Vegetable Clusters in Ethiopia," *European Journal of Development Research* 32 (2020): 1402–1429.

9 ILO (International Labour Organization), "More than 60 Per Cent of the World's Employed Population Are in the Informal Economy" (news release), April 30, 2018.

10 A. Wilkinson et al., "Local Response in Health Emergencies: Key Considerations for Addressing the COVID-19 Pandemic in Informal Urban Settlements," *Environment and Urbanization* 32, no. 2 (2020): 503–522.

11 FAO (Food and Agriculture Organization of the United Nations), *Social Protection and COVID-19 Response in Rural Areas* (Rome: 2020).

12 S. Roy and S. Tripathi, "Migrant Voices: Stories of India's Internal Migrant Workers during the COVID-19 Pandemic," IHRB (Institute for Human Rights and Business), Nov. 24, 2020.

13 C. Afshar and A. Devenish, "Reaching the 'Missing Middle': Social Protection for Informal Workers in COVID-19," *WIEGO Blog*, WIEGO (Women in Informal Employment: Globalizing and Organizing), Oct. 1, 2020.

14 J. Alawa et al., "Addressing COVID-19 in Humanitarian Settings: A Call to Action," *Conflict and Health* 14, no. 64 (2020): 1–4.

15 S. Truelove et al., "The Potential Impact of COVID-19 in Refugee Camps in Bangladesh and Beyond: A Modeling Study," *PLoS Medicine* 17, no. 6 (2020): e1003144.

16 A. Peterman et al., "Pandemics and Violence against Women and Children," CGD Working Paper 528 (Center for Global Development, Washington, DC, 2020).

17 A.R. Quisumbing et al., "Why Gender Matters in COVID-19 Responses: Now and in the Future," in *COVID-19 & Global Food Security*.

18 D. Headey and M. Ruel, "The COVID-19 Nutrition Crisis: What to Expect and How to Protect," in *COVID-19 & Global Food Security*.

19 N. Kumar and A.R. Quisumbing, "Gendered Impacts of the 2007–2008 Food Price Crisis: Evidence Using Panel Data from Rural Ethiopia," *Food Policy* 38 (2013): 11–22.

20 A.R. Quisumbing, N. Kumar, and J.A. Behrman, "Do Shocks Affect Men's and Women's Assets Differently? Evidence from Bangladesh and Uganda," *Development Policy Review* 36, no. 1 (2018): 3–34.

21 N. Kumar et al., "Social Networks, Mobility, and Political Participation: The Potential for Women's Self-Help Groups to Improve Access and Use of Public Entitlement Schemes in India," *World Development* 114 (2019): 28–41.

22 U. Gentilini et al., "Social Protection and Jobs Responses to COVID-19: A Real-Time Review of Country Measures," living paper, December 11, (Washington, DC: World Bank, 2020).

23 Ibid.

24 U. Gentilini, "A Game Changer for Social Protection? Six Reflections on COVID-19 and the Future of Cash Transfers," *Let's Talk Development*, World Bank blog, January 11, 2021.

25 U. Gentilini, S. Khosla, and M. Almenfi, "Cash in the City: Emerging Lessons from Implementing Cash Transfers in Urban Africa," Social Protection & Jobs Discussion Paper No. 2101 (World Bank, Washington, DC, 2021).

26 A. Mukherjee, *Digital Cash Transfers for Stranded Migrants: Lessons from Bihar's COVID-19 Assistance Program*, CGD Note (Washington, DC: Center for Global Development, 2020).

27 World Bank, *Scaling Up Social Assistance Payments as Part of the COVID-19 Pandemic Response* (Washington, DC: 2020).

28 P. Bance and U. Gentilini, "'Suspended until Further Notice': Is There a Role for Public Works in COVID-19 (Coronavirus) Response?," *Jobs and Development Partnerships Blog*, no. 7, Jobs Group, World Bank, April 27, 2020.

29 U. Gentilini et al., *Social Protection and Jobs Responses to COVID-19.*

30 M. Hidrobo et al., *Gender-Sensitive Social Protection: A Critical Component of the COVID-19 Response in Low- and Middle-Income Countries*, IFPRI Issue Brief (Washington, DC: IFPRI, 2020).

31 C. Hawkes, "COVID-19 and the Promise of Food System Innovation," in *COVID-19 & Global Food Security.*

CHAPTER 6

1 D. Laborde et al., "COVID-19 Risks to Global Food Security," *Science* 369, no. 5603 (2020): 500–502.

2 A. Bouët and D. Laborde, "COVID-19 Border Policies Create Problems for African Trade and Economic Pain for Communities," in *COVID-19 and Global Food Security*, ed. J. Swinnen and J. McDermott (Washington, DC: IFPRI, 2020); M. Torero, "Without Food, There Can Be No Exit from the Pandemic: Comment," *Nature* 580 (2020): 588–589.

3 See, for example, A. Winston, *The Big Pivot: Radically Practical Strategies for a Hotter, Scarcer, and More Open World* (Boston: Harvard Business Review Press, 2014).

4 T. Reardon et al., "'Pivoting' by Food Industry Firms to Cope with COVID-19 in Developing Regions: E-commerce and 'Co-pivoting' Intermediaries," *Agricultural Economics* (2021, forthcoming).

5 T. Reardon et al., "Agrifood Industry Transformation and Small Farmers in Developing Countries," *World Development* 37, no. 11 (2009); T. Reardon and C. P. Timmer, "Five Inter-linked Transformations in the Asian Agrifood Economy: Food Security Implications," *Global Food Security* 3, no. 2 (2014): 108–117.

6 K. Van Hoyweghen et al., "The Resilience of Horticultural Supply Chains to the Covid-19 Pandemic: Insights from Senegal," *Agricultural Economics* (2021, forthcoming).

7 Ibid.

8 B. Minten, B. Mohammed, and S. Tamru, "Emerging Medium-Scale Tenant Farming, Gig Economies, and the COVID-19 Disruption: The Case of Commercial Vegetable Clusters in Ethiopia," *European Journal of Development Research* 32 (2020): 1402–1429.

9 T. Reardon, M.F. Bellemare, and D. Zilberman, "How COVID-19 May Disrupt Food Supply Chains in Developing Countries," in *COVID-19 & Global Food Security.*

10 T. Reardon et al., "'Pivoting' by Food Industry Firms."

11 Ibid.

12 Ibid.; "Flipkart Preparing to Start Hyperlocal Delivery Services," ETRetail.com from the *Economic Times*, June 17, 2020.

13 A. Shrivastava, "Udaan Sees a Big Take-Off from Small Cities," *Economic Times*, July 6, 2020; eTail, *Amazon vs Walmart: The Battle for the eCommerce Crown*, Boston Research Briefs (2019); T. Reardon et al., "'Pivoting' by Food Industry Firms."

14 M. Singh, "Indian Online Grocery Startup BigBasket Raises $60M," Techcrunch.com, April 9, 2020; M. Nandy, "Uber Partners with BigBasket to Deliver Essentials to Customers," Livemint, April 2, 2020.

15 T. Reardon et al., "'Pivoting' by Food Industry Firms."

16 K. Chen et al., "How China Can Address Threats to Food and Nutrition Security from the Covid-19 Outbreak," in *COVID-19 & Global Food Security*; S. Fan, W. Si, and Y. Zhang, "How to Prevent a Global Food and Nutrition Security Crisis under COVID-19?," *China Agricultural Economic Review* 12, no. 3 (2020): 471–480.

REGIONAL DEVELOPMENTS

AFRICA

1 Africa CDC (Centres for Disease Control and Prevention), "Coronavirus Disease 2019 (COVID-19)," accessed February 4, 2021, https://africacdc.org/covid-19/.

2 L. Nordling, "Africa's Pandemic Puzzle: Why So Few Cases and Deaths?," *Science* 369, no. 6505 (2020): 756–757; J.M. Maeda and J.N. Nkengasong, "The Puzzle of the COVID-19 Pandemic in Africa," *Science* 371 (6524): 27–28.

3 D. Olewe, "Coronavirus in Africa: Whipping, Shooting and Snooping," BBC News, April 8, 2020.

4 H. Morsy, L. Balma, and A.N. Mukasa, *"Not a Good Time": Economic Impact of COVID-19 in Africa*, Working Paper Series No. 338 (Abidjan: African Development Bank, 2020); IMF (International Monetary Fund), *World Economic Outlook 2020* (Washington, DC: 2020); A.G. Zeufack et al., "Assessing the Economic Impact of Covid-19 and Policy Responses in Sub-Saharan Africa," *Africa's Pulse* 21 (April 2020).

5 Morsy et al., *"Not a Good Time": Economic Impact of COVID-19 in Africa*; Zeufack et al., "Assessing the Economic Impact of Covid-19."

6 Ibid.

7 S. Amewu et al., "The Economic Costs of COVID-19 in Sub-Saharan Africa: Insights from a Simulation Exercise for Ghana," *European Journal of Development Research* 32 (2020): 1353–1378; K.S. Andam et al., "Impacts of COVID-19 on Food Systems and Poverty in Nigeria," in *Advances in Food Security and Sustainability*, vol. 5, ed. M.J. Cohen (Cambridge, MA: Elsevier, 2020); C. Arndt et al., "Covid-19 Lockdowns, Income Distribution, and Food Security: An Analysis for South Africa," *Global Food Security* 26 (2020): 100410.

8 World Bank, "COVID-19 High Frequency Phone Survey 202 (Burkina Faso, 2020)," February 11, 2021; K. Hirvonen, A. de Brauw, and G.T. Abate, "Food Consumption and Food Security during the COVID-19 Pandemic in Addis Ababa," Discussion Paper 1964 (IFPRI, Washington, DC, 2020); M. Amare et al., "Impacts of COVID-19 on Food Security: Panel Data Evidence from Nigeria," IFPRI Discussion Paper 1956 (IFPRI, Washington DC, 2020).

9 D. Resnick, "COVID-19 Lockdowns Threaten Africa's Vital Informal Urban Food Trade," *IFPRI COVID-19 Blog*, March 31, 2020.

10 Hirvonen et al., "Food Consumption and Food Security."

11 A. Tesfaye, Y. Habte, and B. Minten, "COVID-19 Is Shifting Consumption and Disrupting Dairy Value Chains in Ethiopia," *IFPRI COVID-19 Blog*, June 1, 2020.

12 Arndt et al., "Covid-19 Lockdowns, Income Distribution, and Food Security"; Morsy et al., *"Not a Good Time": Economic Impact of COVID-19 in Africa.*

13 Tesfaye et al., "COVID-19 Is Shifting Consumption and Disrupting Dairy Value Chains in Ethiopia"; Hirvonen et al., "Food Consumption and Food Security."

14 IFPRI, COVID-19 Policy Response Portal, accessed 2021.

15 Ibid.

16 U. Gentilini et al., "Social Protection and Jobs Responses to COVID-19: A Real-Time Review of Country Measures," living paper version 12, July 10, 2020 (Washington, DC: World Bank).

17 Ibid.

18 IMF, "Fiscal Monitor Database of Country Fiscal Measures in Response to the COVID-19 Pandemic," October 2020.

19 Ibid.

20 AU (African Union), "The Malabo Declaration on Accelerated Agricultural Growth and Transformation for Shared Prosperity and Improved Livelihoods" (Addis Ababa: African Union Commission, 2014).

21 AUC (African Union Commission), *Second Biennial Review Report of the African Union Commission on the Implementation of the Malabo Declaration on Accelerated Agricultural Growth and Transformation for Shared Prosperity and Improved Livelihoods*, 2019 Progress Report to the Assembly (Addis Ababa: 2020).

22 FEC (Fédérations des Entreprises du Congo), "Enquête sur l'impact de la COVID-19 sur les entreprises du Congo" (Rapport interne, Kinshasa, 2020).

23 E. Giokos, "Across Africa, the Pandemic Reveals Both Inequality and Innovation," CNN, June 12, 2020; WHO (World Health Organization), "WHO Showcases Leading African Innovations in COVID-19 Response," May 21, 2020.

24 B. Bett, D. Randolph, and J. McDermott, "Africa's Growing Risk of Diseases That Spread from Animals to People," *IFPRI COVID-19 Blog*, April 7, 2020.

25 D. Resnick, X. Diao, and G. Tadesse, eds., *Sustaining Africa's Agrifood System Transformation: The Role of Public Policies*, ReSAKSS (Regional Strategic Analysis and Knowledge Support System) 2020 Annual Trends and Outlook Report (Washington, DC, and Kigali: IFPRI and AKADEMIYA2063, 2020).

26 A. Quisumbing et al., "Why Gender Matters in COVID-19 Responses – Now and in the Future," *IFPRI COVID-19 Blog*, April 22, 2020.

MIDDLE EAST AND NORTH AFRICA

1 IFPRI, COVID-19 Policy Response Portal, accessed 2021.

2 C. Breisinger et al., *Impact of COVID-19 on the Egyptian Economy: Economic Sectors, Jobs, and Households*, MENA (Middle East and North Africa) Policy Note 6 (Washington, DC: IFPRI, 2020); M. Raouf, D. Elsabbagh, and M. Wiebelt, *Impact of COVID-19 on the Jordanian Economy: Economic Sectors, Food Systems, and Households*, MENA Policy Note 9 (Washington, DC: IFPRI, 2020).

3 D. Elsabbagh, S. Kurdi, and M. Wiebelt, *Impact of COVID-19 on the Yemeni Economy: Economic Sectors, Food Systems, and Households*, MENA Policy Note 15 (Washington, DC: IFPRI, 2021).

4 Breisinger et al., *Impact of COVID-19 on the Egyptian Economy: Economic Sector, Jobs, and Households*; Raouf et al., *Impact of COVID-19 on the Jordanian Economy.*

5 F. Abdelaziz et al., *Impact of COVID19 on Egypt's Dairy and Artichoke Value Chains*, MENA Regional Program Working Paper (Washington, DC: IFPRI, forthcoming).

6 J.-L. Figueroa, M. Mahmoud, and C. Breisinger, *The Role of Agriculture and Agro-processing for Development in Jordan*, MENA Regional Program Working Paper 5 (Washington, DC, and Cairo: IFPRI, 2018).

7 Abdelaziz et al., *Impact of COVID19 on Egypt's Dairy and Artichoke Value Chains.*

8 Ibid.

9 K.A. Abay and H. Ibrahim, *Winners and Losers from COVID-19: Evidence from Google Search Data for Egypt*, MENA Policy Note 8 (Washington, DC: IFPRI, 2020).

10 Elsabbagh et al., *Impact of COVID-19 on the Yemeni Economy.*

11 D. Headey et al., "Impacts of COVID-19 on Childhood Malnutrition and Nutrition-Related Mortality," *Lancet* 396, no. 10250 (2020): 519–521.

12 C. Breisinger et al., *COVID-19 and the Egyptian Economy: From Reopening to Recovery: Alternative Pathways and Impacts on Sectors, Jobs, and Households*, MENA Policy Note 10 (Washington, DC: IFPRI, 2020).

13 The Sudanese pound (SDG) has sharply depreciated over the past months. Exchange rate is as of Feb. 28, 2021.

14 Abay and Ibrahim, *Winners and Losers from COVID-19.*

15 For more information, see IFPRI, "Agricultural Investment Data Analyzer (AIDA): Guiding Agricultural Investments for Higher Impact," virtual launch event, February 24, 2021.

CENTRAL ASIA

1 For detailed tracking of these policy responses, see IFPRI, COVID-19 Policy Response Portal, updated December 18, 2020.

2 IMF (International Monetary Fund), "Regional Economic Outlook: Middle East and Central Asia," October 2020.

3 Center for Economic Research and Reforms and UNDP (United Nations Development Programme), "Study of the Impact of COVID-19 on Small and Medium-Sized Businesses in the Second Quarter of 2020" (Tashkent: UNDP, 2020).

4 Ibid.

5 Data for January-September 2020, Kazakhstan, Bureau of National Statistics database, accessed December 14, 2020; Data for January-November 2020: Kyrgyzstan, National Statistical Committee, "Gross Domestic Product, January-November 2020," accessed Dec. 10, 2020; Data for January-September 2020: Uzbekistan, State Committee on Statistics, "Uzbekistan by the Numbers (as of January-September 2020)."

6 FAO, *Europe and Central Asia: Regional Food Market Situation and Policy Bulletin in Response to the COVID-19 Pandemic*, Issue 1 (April 29, 2020).

7 National statistical offices of respective countries, see note 5.

8 ADB (Asian Development Bank), *Asian Development Outlook Supplement*, December (Manila: 2020); ADB, *Asian Development Outlook 2020 Update: Wellness in Worrying Times*, September 2020 (Manila: 2020).

9 IMF, IMF Primary Commodity Price System, accessed Jan. 28, 2021.

10 Calculated using data from National Bank of Kazakhstan, "Balance of Payments," Dec. 30, 2020.

11 Calculated using data from Uzbekistan, State Committee on Statistics, "Foreign Economic Activity," accessed Jan. 7, 2021.

12 Calculated using data from National Bank of Kazakhstan, "Balance of Payments"; National Bank of the Kyrgyz Republic, "Balance of Payments of the Kyrgyz Republic," accessed Jan. 7, 2021; and Uzbekistan, State Committee on Statistics, "Foreign Economic Activity."

13 E. Asfaw, M. Li, I. Rudenko, and F. Saydullaeva, *Urban Food Security and COVID-19 Crisis: The Case of Uzbekistan Cities*, ECFS Case Study (Moscow: Eurasian Center for Food Security, forthcoming, 2021).

14 ITU (International Telecommunication Union), "Economic Impact of COVID 19 on Digital Infrastructure," Report of Economic Experts Roundtable organized by ITU, GSR-20 Discussion Paper, July 2020.

15 World Bank, "COVID-19 Prompts Urgency of Bridging Digital Divide in Central Asia" (press release), Almaty, Dec. 2, 2020.

16 IFPRI's phone survey in Southern Tajikistan was conducted in September-October 2020. The survey sample size was 1,200; of these, 1,150 households had been part of the sample of a 2018 IFPRI survey, which

allowed us to create a household panel. For more details, see K. Akramov and K. Kosec, "Impact of the COVID-19 Pandemic on Household Welfare, Food Security, and Agriculture in Central Asia," Powerpoint, presented at the Virtual Policy Seminar on Food Policy Research and Capacity Development in Eurasia organized by Eurasian Center for Food Security (ECFS) at Lomonosov Moscow State University, the World Bank, and IFPRI, December 2, 2020.

17 UNDP Uzbekistan, "Applying Data on COVID-19's National Impacts to Families and Businesses" (press release), November 11, 2020.

18 Akramov and Kosec, "Impact of the COVID-19 Pandemic on Household Welfare, Food Security, and Agriculture in Central Asia."

19 IMF, "Regional Economic Outlook: Middle East and Central Asia."

20 C. Lakhner et al., "Updated Estimates of the Impact of COVID-19 on Global Poverty: Looking Back at 2020 and the Outlook for 2021," *Data Blog*, World Bank, Jan. 11, 2021.

SOUTH ASIA

1 L.L. Song, "How South Asia Can Continue as World's Fastest Growing Subregion," op-ed, Asian Development Bank, August 2, 2019.

2 See S. Rashid, A. Ahmed, and A.W. Rana, "South Asia," in *2020 Global Food Policy Report* (Washington, DC: IFPRI, 2020).

3 These estimates do not include the Maldives and Bhutan.

4 World Bank, "Real GDP Growth at Constant Market Prices, 2019 and Forecasts," accessed March 1, 2021.

5 These estimates are compiled by the authors from IMF data and IFPRI's COVID-19 Policy Response Portal.

6 For counter arguments regarding social distancing and lockdowns, see A.M. Mobarak, "Responding to COVID-19 in the Developing World," *Yale Insights*, April 3, 2020.

7 This caused serious challenges with social distancing and led to the late summer surge.

8 Supports to agriculture came in the form of subsidized credits, distribution of subsidized inputs, and increased efforts in public procurement. Both India and Bangladesh allocated over 15 percent of total COVID-19 response budgets to agriculture.

9 See D. Varshney et al., "COVID-19, Government Transfer Payments, and Investment Decisions in Farming Business: Evidence from Northern India," *Applied Economic Perspectives and Policy* 43, no. 1 (2021): 248-269, for pandemic and agriculture; S. Narayanan, "How India's Agrifood Supply Chains Fared during the COVID-19 Lockdown, from Farm to Fork," *IFPRI COVID-19 Blog*, July 20, 2020, for supply chain and food prices; and S. Chowdhury, "Op-Ed: A Lifeline against Hunger," *Dhaka Tribune*, December 26, 2020, for operation of the largest safety net program in Bangladesh during the pandemic.

10 The World Bank's South Asia website has this statement: "South Asia's economic outlook is grim. The region is set to plunge in 2020 into its worst-ever recession as the devastating impacts of COVID-19 on South Asian economies linger on" (World Bank, "South Asia: Overview," October 8, 2020).

11 Varshney et al., "COVID-19, Government Transfer Payments, and Investment Decisions in Farming Business: Evidence from Northern India."

12 This study was conducted by IFPRI's POSHAN program; results cannot be reported here because the study is under review for publication.

13 S. Balsari, M. Sange, and Z. Udwadia, "Comment: Covid-19 Care in India: The Course to Self-Reliance," *Lancet*, published online Aug. 25, 2020.

14 "PM to Launch Disbursement of Cash Aid to 50 Lakh Families Tomorrow," *Business Standard*, May 13, 2020.

15 H. Rahman et al., *Livelihoods, Coping and Support during Covid-19 Crisis* (Dhaka: Power and Participation Research Centre and BRAC Institute of Governance and Development, 2020).

16 These results will be published soon.

17 M.S. Moeen et al., "Estimating the Economic Impacts of the First Wave of COVID-19 in Pakistan Using a SAM Multiplier Model," IFPRI Discussion Paper 2001 (IFPRI, Washington, DC, 2021).

18 Pakistan Bureau of Statistics, "Key Findings of Survey for Evaluating the Socio-economic Impact of COVID-19" (PowerPoint presentation, 2021).

19 H. Alderman, J. Hoddinott, and B. Kinsey, "Long Term Consequences of Early Childhood Malnutrition," *Oxford Economic Papers* 58, no. 3 (2006): 450-474; J. Hoddinott and B. Kinsey, "Child Growth in the Time of Drought," *Oxford Bulletin of Economics and Statistics* 63, no. 4 (2001): 409-436; J. Hoddinott, "Shocks and Their Consequences across and within Households in Rural Zimbabwe," *Journal of Development Studies* 42, no. 2 (2006): 301-321.

20 For South Asia-specific information, see F. Dizon and A. Herforth, "The Cost of Nutritious Food in South Asia," Policy Research Working Paper 8557 (World Bank, Washington, DC, 2018).

21 K. Raghunathan, D. Headey, and A. Herforth, "Affordability of Nutritious Diets in Rural India," IFPRI Discussion Paper 1912 (IFPRI, Washington, DC, 2020).

22 Early results from an IFPRI survey suggest that cutting food consumption was one of the major coping strategies of the poor, which also impacted dietary patterns, as they had to consume more low-cost cereals. See S.K. Chowdhury et al., "Public Food Transfers during Pandemic: Insights from an IFPRI Survey in Bangladesh," *Financial Express*, Dec. 22, 2020. Although not yet published, a more comprehensive IFPRI-Cornell University survey in Bangladesh points in the same direction.

23 For projections, see World Bank, "World Bank Predicts Sharpest Decline of Remittances in Recent History," press release, April 22, 2020.

24 Recent estimates suggest that, although remittance inflows declined for the region, they increased in Bangladesh and Pakistan.

25 Leakages in these programs are found to be high in all countries. See S. Rashid, A. Gulati, and R. Cummings Jr., *From Parastatals to Private Trade: Lessons from Asian Agriculture* (Washington, DC: IFPRI, 2008). In Bangladesh, where a new program, called the Food Friendly Program (FFP), was functioning much better than similar earlier programs, an IFPRI study during the pandemic found that leakage went up by more than 15 percent compared to 2018. See S. Chowdhury et al., *Public Food Transfers during a Pandemic: Insights from Bangladesh*, IFPRI Policy Brief (Washington, DC: IFPRI, 2021).

26 The policies, such as minimum support price, evolved to promote Green Revolution technologies, and the development research community has long debated about the relevance and unsustainably high costs of those policies. See Rashid, Gulati, and Cummings Jr., *From Parastatals to Private Trade: Lessons from Asian Agriculture*.

EAST AND SOUTHEAST ASIA

1 ADB (Asian Development Bank), *An Updated Assessment of the Economic Impact of COVID-19*, ADB Briefs No. 133 (Manila: 2020); World Bank, *From Containment to Recovery*, East Asia and Pacific Economic Update, October (Washington, DC: 2020).

2 United Nations, *Policy Brief: The Impact of COVID-19 on South-East Asia*, July 2020.

3 World Bank, *From Containment to Recovery*.

4 World Bank, *Global Economic Prospects*, January 2021 (Washington, DC).

5 World Bank, *COVID-19 Crisis through a Migration Lens* (Washington, DC: 2020).

6 X. Diao and K. Mahrt, *Assessing the Impact on Household Incomes and Poverty of Declines in Remittances Due to COVID-19*, Myanmar SSP Policy Note 6 (Washington, DC: IFPRI, 2020).

7 World Bank, *From Containment to Recovery.*

8 S. Rozelle, H. Rahimi, H. Wang, and E. Dill, "Lockdowns Are Protecting China's Rural Families from COVID-19, but the Economic Burden Is Heavy," in *COVID-19 & Global Food Security*, ed. J. Swinnen and J. McDermott (Washington, DC: IFPRI, 2020).

9 WFP (World Food Programme), "COVID-19 Will Double Number of People Facing Food Crises unless Swift Action Is Taken," news release, April 21, 2020.

10 IFPRI, *COVID-19 Policy Response Portal* (Washington, DC: 2020).

11 FAO (Food and Agriculture Organization of the United Bations), *Ample Supplies to Help Shield Food Markets from the COVID-19 Crisis* (Rome: 2020).

12 "Philippines Typhoon Recovery, Complicated by Coronavirus Concerns," UN News, May 15, 2020.

13 K.Z. Chen, S. Fan, and Y. Zhan, "COVID-19 and Food Security: Early Responses, Impact and Lessons from China," editorial, *China Agricultural Economic Review* 12: 381–385; Y. Zhan and K.Z. Chen, "Building Resilient Food System Amidst COVID-19: Reponses and Lessons from China," *Agricultural Systems* 190 (2021).

14 P. Arunmas, "Ministry Creates War Rooms to Handle Essential Goods," *Bangkok Post*, March 26, 2020.

15 "Food Delivery on the Rise in ASEAN," *ASEAN Post*, May 4, 2020.

16 Y. Zhang, X. Diao, K.Z. Chen, S. Robinson, and S. Fan, "Impact of COVID-19 on China's Macroeconomy and Agri-food System: An Economy-wide Multiplier Model Analysis," *China Agricultural Economic Review* 12, no. 3 (2020): 387–407.

17 FAO, *Impacts of Coronavirus on Food Security and Nutrition in Asia and the Pacific: Building More Resilient Food Systems* (Rome: 2020).

18 World Bank, *From Containment to Recovery.*

19 China, State Council Information Office, "China–ASEAN Trade Grows Robustly Despite COVID-19 Pandemic," Aug. 28, 2020.

20 K.Z. Chen and R. Mao, "Fire Lines as Fault Lines: Increased Trade Barriers during the COVID-19 Pandemic Further Shatter the Global Food System," *Food Security* 12 (2020): 735–738.

21 ASEAN, "Statement of ASEAN Ministers on Agriculture and Forestry in Response to the Outbreak of the Coronavirus Disease (Covid-19) to Ensure Food Security, Food Safety and Nutrition in ASEAN," April 15, 2020.

22 ADB, *Regional Comprehensive Economic Partnership: Overview and Economic Impact*, ADB Briefs No. 164 (Manila: 2020).

23 ADB, *Food Security in Asia and the Pacific amid the COVID-19 Pandemic*, ADB Briefs No. 139 (Manila: 2020).

LATIN AMERICA AND THE CARIBBEAN

1 Johns Hopkins University, CSSE Database, https://coronavirus.jhu.edu/data/mortality.

2 IMF (International Monetary Fund), *Regional Economic Outlook for Western Hemisphere*, October (Washington: 2020).

3 E. Díaz-Bonilla and F. Paz, "Food in the Time of Pandemic: Realism without Magic," *IFPRI Blog: Event Post,* June 19, 2020.

4 IMF, "Fiscal Monitor Database of Country Fiscal Measures in Response to the COVID-19 Pandemic," IMF Fiscal Affairs Department, October 2020.

5 E. Díaz-Bonilla, D. Laborde, and V. Piñeiro, *Covid-19: The Impact on Food Security in Latin America and the Caribbean* (Washington, DC: InterAmerican Development Bank, forthcoming); MIRAGRODEP Model, http://www.agrodep.org/model/miragrodep-model.

6 A. Herforth et al., "Cost and Affordability of Healthy Diets across and within Countries," background paper for the *State of Food Security and Nutrition in the World 2020* (FAO, Rome, 2020).

7 Nonweighted average of the countries in the World Bank's ASPIRE database (https://www.worldbank.org/en/data/datatopics/aspire).

8 E. Díaz-Bonilla, "Fiscal and Monetary Responses to the COVID-19 Pandemic: Some Thoughts for Developing Countries and the International Community," in *COVID-19 & Global Food Security*, ed. J. Swinnen and J. McDermott (Washington, DC: IFPRI, 2020).

9 E. Díaz-Bonilla, "América Latina y el Caribe en la economía verde y azul," *Pensamiento Iberoamericano* 7 (2019): 94–105.

Made in the USA
Monee, IL
05 November 2026

67387144R00074